Geoheritage, Geoparks and Geotourism

Conservation and Management Series

Spectacular geo-morphological landscapes and regions with special geological features or mining sites are becoming increasingly recognized as critical areas to protect and conserve for the unique geoscientific aspects they represent and as places to enjoy and learn about the science and history of our planet. More and more national and international stakeholders are engaged in projects related to "Geoheritage", "Geo-conservation", "Geoparks" and "Geotourism"; and are positively influencing the general perception of modern Earth Sciences. Most notably, "Geoparks" have proven to be excellent tools to educate the public about Earth Sciences; and they are also important areas for recreation and significant sustainable economic development through geotourism. In order to develop further the understanding of Earth Sciences in general and to elucidate the importance of Earth Sciences for Society, the "Geoheritage, Geoparks and Geotourism Conservation and Management Series" has been launched together with its sister "GeoGuides" series. Projects developed in partnership with UNESCO, World Heritage and Global Geoparks Networks, IUGS and IGU, as well as with the 'Earth Science Matters' Foundation will be considered for publication. This series aims to provide a place for in-depth presentations of developmental and management issues related to Geoheritage and Geotourism in existing and potential Geoparks. Individually authored monographs as well as edited volumes and conference proceedings are welcome; and this book series is considered to be complementary to the Springer-Journal "Geoheritage".

More information about this series at http://www.springer.com/series/11639

Abhik Chakraborty • Kuniyasu Mokudai •
Malcolm Cooper • Mahito Watanabe •
Shamik Chakraborty
Editors

Natural Heritage of Japan

Geological, Geomorphological, and Ecological Aspects

Springer

Editors
Abhik Chakraborty
Center for Tourism Research (CTR)
Wakayama University
Wakayama, Japan

Kuniyasu Mokudai
Japan Geo Service Inc.
Tokyo, Japan

Malcolm Cooper
College of Asia Pacific Studies
Ritsumeikan Asia Pacific University
Beppu, Oita, Japan

Mahito Watanabe
Geological Survey of Japan
National Institute of Advanced Industrial Science and Technology (AIST)
Tsukuba, Japan

Shamik Chakraborty
Institute for the Advanced Study of Sustainability
United Nations University
Tokyo, Japan

ISSN 2363-765X ISSN 2363-7668 (electronic)
Geoheritage, Geoparks and Geotourism
ISBN 978-3-319-61895-1 ISBN 978-3-319-61896-8 (eBook)
DOI 10.1007/978-3-319-61896-8

Library of Congress Control Number: 2017951396

Printed on acid-free paper

This Springer imprint is published by Springer Nature
The registered company is Springer International Publishing AG
The registered company address is: Gewerbestrasse 11, 6330 Cham, Switzerland

Foreword

When Europeans think about Japan, they think mainly about its culture—seen as an exotic culture embracing temples, samurai, bonsai, sumo wrestling, kimonos, bullet trains (shinkansen), cars and motorbikes, cameras and televisions, and Japanese food (for an increasing number of Japanese restaurants have opened up in European cities). They rarely think about Japanese nature though, of course, they do know about Mt. Fuji, and in Britain at least, they were introduced to *Japan: the enchanted islands*, a series of three BBC programs that focused on nature and its human interactions on Honshu, the South Western islands, and Hokkaido.

My own knowledge of Japan was similarly restricted, so when I was invited to give a presentation on geoconservation at the 2015 Japanese Geoscience Union (JpGU) conference in Makuhari, I jumped at the chance to visit the country and learn more about its nature and, indeed, its culture. The invitation had come about because one of the editors of this book, Abhik Chakraborty, had read my book *Geodiversity: valuing and conserving abiotic nature* (Gray 2013) and wanted me to input to a session on geoconservation at the conference.

The Visit

With me in Japan was my partner Pauline, and we started our visit in the cultural capital of Japan, Kyoto. Here we visited many temples and their gardens including Arashiyama, Ryoanji, Kinkakuji (golden pavilion), Ginkakuji (silver pavilion), Nanzenji, Chion-in, and Kiyomizudera. The most interesting point about the temple gardens was the way in which stone was an important part of the garden design and formed an integrated whole along with the plants and water. And at Ryoanji, it was amazing to witness the scores of visitors contemplating the Zen rock garden with its larger stones set amidst a sea of smaller pebbles raked into sweeping linear patterns that are said to facilitate meditation. Here was a great example of an ancient use of stone and a modern example of geotourism. While in Kyoto, we took a day trip to Nara where we visited the equally stunning gardens of Yoshiki-en and Isui-en as well as the temples of Kofuku-ji and Todai-ji and the Shinto shrine of Kasuga Taisha, meeting many of Nara's famous deer along the way.

After Kyoto, we moved on to Takayama from where we visited the beautiful and tranquil valley at Kamikochi in the Japanese Alps. It also lies within Chubusangaku National Park, one of 34 national parks in Japan. The valley was surrounded by impressive snow-covered peaks, while the walk beside the braided Azusa River was an absolute delight.

Then to Makuhari/Tokyo where I gave three different geodiversity/geoconservation presentations to the Japanese Geopark Network, the Tokyo Geographical Society, and the Japanese Geoscience Union (JpGU2015). In these three lectures, I tried to outline the following points:

1. Geodiversity is the abiotic equivalent of biodiversity and can be defined as "the natural range (diversity) of geological (rocks, minerals, fossils), geomorphological (landforms, topography, physical processes), soil and hydrological features. It includes their assemblages, structures, systems and contributions to landscapes" (Gray 2013).
2. Geoheritage is the collective term for those specific features of geodiversity that may be identified as being worthy of geoconservation. This may be because of their scientific or other values.
3. Just as biodiversity provides human societies with many "ecosystem services" or "natural services" (Daily 1997), geodiversity brings a huge variety of goods and services that benefit us in our everyday lives. In fact our modern, complex society could not function without the planet's geodiversity, which has been brilliantly exploited by generations of humans from the Stone Age to the Silicon Age. Some sites are of scientific value in giving us the evidence for the history of the planet/local area or the evolution of life, others provide attractive natural or seminatural landscapes, and many provide the variety of physical habitats in which biodiversity has evolved. Indeed, as part of a theatrical metaphor, geodiversity is now often seen as the stage upon which the wildlife actors can thrive (Beier, Hunter, and Anderson 2015).
4. Geodiversity and geoheritage can be threatened by many types of human activities including mineral extraction that can destroy landforms, land development and urban expansion, coastal and river engineering which can interrupt the operation of natural processes, forestry and agriculture, recreation and tourism, military operations, and overcollecting of fossils and minerals by commercial operations or by geoscientists themselves.
5. Because geodiversity and geoheritage are of value to us in various ways yet may be lost, damaged, or polluted by human activities, there is an unanswerable case for geoconservation. This can be achieved by various methods that include site management, curation, licensing of field visits and research, site supervision, benevolent ownership, restoration, legislation, policy development, and, most importantly, education.
6. Geodiversity is the backbone of geoheritage conservation as many countries, including the USA, the UK, Ireland, and New Zealand, try to conserve sites that are representative of the nation's geodiversity.

After the conference, we were able to visit both the Hakusan Tedorigawa and the Izu Peninsula national geoparks and I was very impressed by the way in which Japan has so strongly adopted the geopark initiative. Geoparks now exist all over Japan, with several being recognized as UNESCO Global Geoparks. They have been established in a variety of geological/geomorphological environments, one of the main drivers being a desire to increase geotourism to these places and thus to stimulate the local economies. At JpGU2015, the geopark sessions were among the best attended in the whole conference, reflecting the interest in promoting geotourism based on the local geodiversity.

On the last day of our visit to Japan, our request was to visit Mt. Fuji and we were kindly driven up to the 5th station on the south side of the cone. This visit to Japan's most famous natural attraction did not disappoint and made a memorable end to our visit to Japan.

Reflections

This was a very limited first visit to Japan, taking us only to parts of central Honshu. Clearly, the natural features of the Japanese archipelago are much more diverse. But we now have had an introduction to Japanese nature and hope to extend this to other parts of the country some day. Let me finish this foreword with two main reflections.

First, one of the main aims of geoconservation is to allow natural, dynamic processes to continue operating, but in Japan, and indeed in other tectonically active areas of the world, this poses a major challenge for geoconservation, given the hazardous nature of these processes and the very real threats they pose to life and property. Some of these natural hazards—earthquakes, volcanic eruptions, and tsunamis—are impossible to control at source, and so the response is to try to predict their occurrence, issue warnings, and evacuate populations. In fact, we experienced no less than two earthquakes in the 3 weeks we were in Japan. But in the case of other natural or seminatural hazards—e.g., slope instability, flooding, and coastal erosion—the understandable response has been to engineer the land to control these natural processes and restrict their operation. This has had a huge impact on the natural landscapes of Japan and perhaps needs some reassessment. Could some, especially redundant constructions, be removed and more natural landscapes restored? Could measures be designed to be less prominent in the landscape?

Secondly, although Japan's tectonic and volcanic history has created problems for its population and not provided it with a rich supply of geological resources, it has brought many benefits. First, Japan has an incredible landscape of volcanoes, mountains, forests, rivers, and rugged coastlines that have inspired artists and writers throughout its long history and are increasingly attracting international tourists. Secondly, its dynamic environment has meant the emergence of international scientific expertise in natural hazard understanding and prediction. Thirdly, its rich volcanic soils have allowed an agricultural landscape dominated by rice cultivation to be developed. Fourthly, Japan's geothermal energy has been tapped by its popular onsens and can potentially be used for geothermal power. Fifthly, as an island nation, Japan has developed strong connections with the sea and its resources. And finally, a rich culture has evolved over generations within and of these landscapes.

This book provides many detailed chapters on the natural landscapes and resources of Japan, both biotic and abiotic. I hope it increases public understanding within Japan of the country's natural environment and, in turn, influences scholars, planners, decision-makers, and the general public to give full attention to the issues raised herein and the need to further conserve Japan's amazing natural heritage.

Queen Mary University of London
London, United Kingdom
2016

Murray Gray

References

Beier P, Hunter ML, Anderson M (2015) Conserving nature's stage. *Conservation Biology*, 29, 613–617.

Daily GC (1997) *Nature's services: societal dependence on natural ecosystems.* Washington DC, Island Press.

Gray M (2013) *Geodiversity: valuing and conserving abiotic nature.* 2nd edition. Chichester, UK, Wiley Blackwell.

Preface

The idea for this volume germinated in an interdisciplinary session at the Japan Geoscience Union Conference in 2015 that was organized by two of the editors of this volume. That session was mainly focused on geoconservation and sustainable development, but eventually we recognized that geoheritage and its conservation is only part of the story, albeit one of critical importance, in finding out how to evaluate our natural heritage and see it as a fundamental asset for future societies. It was also felt that while conservation science so far has focused on key species and the biotic environment, a complementary focus on the planet's abiotic diversity and the importance of earth processes that sustain all biotic diversity should be provided by combining accounts of geological, geomorphological, and ecological heritage in one volume. In that sense, this volume is a unique account of the insights of our writers, who have approached the task of analyzing natural heritage from the geological, ecological, and social sciences. The stage for exploring this multifaceted concept of natural heritage is the Japanese Islands, a region that is known for vigorous earth processes. Several of the natural heritage landscapes here are important enough to have received international heritage branding as UNESCO World Heritage and Global Geoparks, while the National Parks roster also features several heritage areas of outstanding value. Most of these areas are also popular tourist destinations. While visiting such sites can foster the understanding of nature's mechanisms and elevate awareness of nature conservation, tourism can create further pressure on the integrity of ecosystems and endangered species. It is hoped that this volume will be helpful for both the enthusiastic visitor and the prudent planner, both of whom are integral to the maintenance of the value of natural heritage.

This volume has been a journey in itself; it evolved in both its scope and contents, ending up with a slightly different set of chapters than those we originally had in mind. This journey is a reflection of the learning process we underwent—our close look at natural heritage showed us different facets that were obscure when we first planned the book—and this compelled us to include those aspects. At the same time, however, we had to choose between several contending sites for some of the case studies. This volume is by no means therefore an exhaustive account of the natural heritage of Japan, and we hope that it will stimulate further accounts of other heritage sites and landscapes. Of course, the multifaceted evaluation of natural heritage is not a concept that is limited to Japan, and complementary narratives will surely emerge from other parts of the world.

We remain indebted to the authors who contributed to this volume. The first editor wishes to thank Murray Gray for contributing the Foreword, and Koji Wakita of Yamaguchi University and Setsuya Nakada of the University of Tokyo for their contributed chapters despite them being set a very short timeline. We also sincerely thank the Springer team, notably Prasanna Kumar Narayanasamy, Janet Steritt-Brunner, Ravi Vengadachalam, and Annett Buettner, for their cooperation at various stages of this manuscript and for their patience throughout this project.

Wakayama — Abhik Chakraborty
Tokyo — Kuniyasu Mokudai
Beppu — Malcolm Cooper
Tsukuba — Mahito Watanabe
Tokyo — Shamik Chakraborty
2017

Contents

Introduction 1

Abhik Chakraborty

Abstract

This chapter provides an outline of the concept of natural heritage and discusses the main geological, geomorphological, and ecological characteristics of natural heritage of Japan. The chapter begins with referring to the Anthropocene concept and examines how pervasive anthropogenic influence has affected the contemporary geo-biosphere. However, even as our actions have transformed our planet at its surface level, we remain far from replicating the most fundamental earth processes, and in this sense, understanding how natural processes engender our planet's heritage provides important insights for a better future of humanity. Accordingly, the chapter makes the argument that the term "natural heritage" must be reconceptualized through the appreciation of the mechanisms of our dynamic planet. The chapter then provides a brief discussion of the geological history of the Japanese Islands and introduces the main geomorphic and ecological characteristics of the archipelago, in order to explain the setting and scope of this book. By referring to the birth of this island arc in deep time, the chapter explains how the defining characteristics of land formation and landscaping in the Japanese Islands provide a snapshot of the beating heart of our planet. Finally, the chapter closes with an outline of the structure of this volume.

Keywords

Natural heritage • Anthropocene • Japanese Islands • Dynamic heritage

1.1 The Scope of "Natural Heritage" in this Book

1.1.1 Natural Heritage: Why Now?

The words "Natural Heritage" appear frequently in the conversations around us. A simple Google search will yield a large number of hits, these words are regularly featured in the news media, and they appear as topics of several books. In 2006, UNESCO published "The UNESCO World Heritage Centre's Natural Heritage Strategy" (UNESCO 2006), which referred to the following definition of "natural heritage" adopted by its World Heritage Convention:

> . . .natural features consisting of physical and biological formations or groups of such formations, which are of outstanding universal value from the aesthetic or scientific point of view; geological and physiographical formations and precisely delineated areas which constitute the habitat of threatened species of animals and plants of outstanding universal value from the point of view of science or conservation; natural sites or precisely delineated natural areas of outstanding universal value from the point of view of science, conservation or natural beauty.

This broad and comprehensive definition tells us that natural heritage can be of both abiotic and biotic nature; it is usually related to the planet's history—notably the history of land formation, landscaping, and the evolution of life—and it can have scientific or aesthetic value, or both. The

A. Chakraborty (✉)
Center for Tourism Research (CTR), Wakayama University, Wakayama, Japan
e-mail: abhik78@hotmail.co.jp

A. Chakraborty et al. (eds.), *Natural Heritage of Japan*, Geoheritage, Geoparks and Geotourism,
DOI 10.1007/978-3-319-61896-8_1

definition also mentions the word "threatened," which perhaps constitutes the most important underpinning of the various contemporary works on the natural heritage concept. The idea that binds works in contemporary natural heritage management, ecology, and environmental science is that overwhelmingly disproportionate anthropogenic impact on the planet's features and systems has put the future of many species in peril. It was mentioned in a recent BBC report that nearly half of all designated natural heritage sites are affected by poaching, illegal logging, and overfishing (BBC 2017). Crutzen and Stoermer (2000) famously coined the term "Anthropocene" to describe the contemporary era—a distinct geological period that bears its name due to the pervading human presence in the geosphere and our species' ability to change planetary processes—a development that has left an indelible signature on the geological record.

A recent study published in *Science* by the Anthropocene Working Group has presented the evidence of functionally and stratigraphically distinct properties of the Anthropocene (Waters et al. 2016). Though the precise point where this period began is a matter of considerable controversy and even more fundamental questions on whether the Anthropocene concept can be independently verified as a distinct geological period remain (Malm and Hornborg 2014; Autin and Holbrook 2012), it can be argued that large-scale and long-term changes in the natural environment today are integrally related with human agency. There are two international academic journals, "Anthropocene" and "The Anthropocene Review" that provide space for academic debate on this subject. Even natural hazards such as large earthquakes, tsunami, and volcanic eruptions, which seemingly occur due to processes entirely unrelated to the human domain (Clark 2011), leave their impacts through ways that are devised or experienced by the human agency.

An obvious question arises: if earth processes are increasingly being determined by the human agency, has "nature" as understood in the standard definition of the word (i.e., "nature" that represents features, forces, and biota that exist independently of humans) ceased to exist? Castree (2005), in his authoritative work on the concept of "nature," describes how our sciences, philosophies, and thought have grappled with the subject over the years. Wohl (2013) has reviewed how wilderness has effectively disappeared from the face of the planet as nearly all landform and landscape features have undergone some extent of human manipulation. Even on landscapes where human beings are not conspicuous actors, human-induced change can be far-reaching, as evidenced by the melting of Antarctic ice (Rignot and Jacobs 2002; Payne et al. 2004) due to a rise in global temperatures that is at least partially driven by anthropogenic carbon missions.

Yet humans have not replaced nature and remain far from replacing its key mechanisms. Aside from a few very crude and rudimentary examples, humans have not been able to replicate life and its complex structure. We are even farther away from replicating a landform or a landscape; our current engineering capacity is woefully short of altering long-term geologic and geomorphic processes such as crustal formation, mountain building, oceanic currents, and fluvial mechanisms. Nature therefore remains independent of us at those scales, although its mechanisms are being increasingly disturbed by the human agency in a way that affects life, which includes that of our own species. This requires us to take a look around us, to re-appreciate nature in the landforms, landscapes, and ecosystems that have evolved over millennia and marvel at the astonishing complexity and the resilient message that we have to learn from nature rather than try to absorb it within the human agency.

Nature is definitely under threat when seen from a biocentric and, perhaps more, from an anthropocentric point of view. This threat is substantial, and it has received substantial attention across the globe in recent years. However, most of the spotlight is on the loss of biotic nature, on the loss of many species that are valuable to us due to one reason or the other. This has resulted in important gains in conservation of the biodiversity and international commitments and frameworks such as the Convention on Biological Diversity (CBD) to protect it. Climate change has been an equally successful concept in terms of capturing our attention. Unfortunately, understanding of nature's abiotic components and mechanisms is poorer, and there is no convention to protect our planet's abiotic diversity. Gregory and Lewin (2014) point out that geomorphic literacy has not progressed in the manner of climate literacy due to the fact that geomorphic processes operate over very large scales that go beyond human perception. Gray (2013) has presented us with the important framework of "geodiversity"—the diversity of rocks, minerals, fossils, landforms, topography, physical processes, and their relationship with landscapes—and pointed out that this diversity has a fundamental and intrinsic importance. Fortunately, we do have a long history of conservation of natural heritage, as evidenced by the establishment of national parks and reserves in different countries, and international initiatives such as the World Heritage Convention, Biosphere Reserves, the Ramsar Wetland Convention, and most recently, the UNESCO Geoparks Program (International Geoscience and Geoparks Program). Conservation of natural diversity has a profound importance for the well-being of the human society because this diversity nurtures our intellect: our sciences, our arts, our philosophies. As the economies of the nations are ever closely bound together by the movement of information and people, "natural heritage" has a very important economic potential. Several decades ago, the Club of Rome Report (1972) called for transforming our economies from economies of resource exploitation to economies of resource conservation. The message was ahead of its time at that point and generated much controversy. However, its value has not diminished, and many of today's sustainability scholars are aware of the continued relevance of this insight. In this regard, responsible and informed tourism can play a major role as an alternative development pathway. This is why "natural heritage" continues to be an extremely important topic for our contemporary society.

1.1.2 A Dynamic Heritage in a Dynamic Part of the Planet

Three key facets of natural heritage can be summed up in the following manner: (1) they are dynamic, (2) they are complex, and (3) they possess an innate ability to change non-linearly. Influential works in contemporary ecology such as Gunderson and Holling (2001), Abel et al. (2006), and Gunderson et al. (2010) have shown that natural systems can suddenly shift "states" through events that appear as chaotic turmoil on the landscape. If we take a long-term perspective, we may be able to observe that such chaotic events and transitions are fundamental to the integrity of the landscape and ecosystems (Fig. 1.1). This concept is also applicable to geologic and geomorphological systems, albeit those systems might operate across much larger spatial and temporal scales. Scientific knowledge has often been used by the society in order to mitigate the uncertainty posed by the chaotic nature of earth systems, but in doing so we often come up with solutions that are detrimental to the integrity of our planet's heritage. In this book, we see nonlinearity and chaotic change as integral parts of natural heritage.

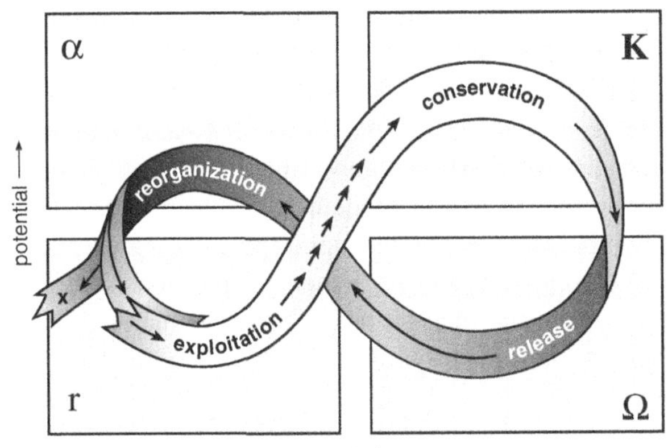

Fig. 1.1 The "adaptive cycle" in the ecological landscape (after Gunderson and Holling 2001) From Panarchy by Lance Gunderson and C.S. Holling. Copyright © 2002 Island Press. Reproduced by permission of Island Press, Washington, DC. The "adaptive cycle" is characterized by a slow buildup of energy followed by a rapid release phase which may appear chaotic but, in essence, is fundamental to growth and renewal in natural systems

Finally, this book is an attempt to present the complexity, attractions, and the efforts and challenges of conserving dynamic and multidimensional natural heritage in a specific part of the world. The Japanese Islands have the distinction of being among the most seismically and volcanically active zones of the planet, and the main islands are intensely inhabited. For this reason, this region offers a particularly valuable lesson for understanding our relationship with natural heritage. The devastation by the East Japan Earthquake and Tsunami is probably still a lingering image of the interaction between nature and society in contemporary Japan. But Japan also offers astounding landforms created out of colossal geological forces and beautiful landscapes created through interactions between surface relief, geomorphological processes, and biodiversity. The dense population and resource-intensive modern lifestyle present unique challenges for conservation of natural heritage in the archipelago, but Japan's signature heritage sites are places of great natural beauty and richness.

1.2 Dynamic Archipelago: Japan's Geological, Geomorphological, and Ecological Characteristics as Heritage

The Japanese Islands are located at the intersection of multiple tectonic plates. The biota and cultural life in these islands are shaped by ongoing geological processes as well as by oceanic and atmospheric circulation patterns. Together, these aspects form the defining contours of Japan's natural heritage.

Geologically, Japan is located at the intersection of at least four large plates: the Eurasian and North American Continental Plates are slowly drifting closer, while the Pacific Plate and the Philippine Sea Plates are subducting at the continental margin. The birth of Japan is thought to have taken place at the rifted continental margin of the Rodinia Supercontinent approximately 750–700 Ma (Maruyama et al. 1997). The breakup of Rodinia is linked to the opening of the Proto-Pacific Ocean (Panthalassa) which began subducting at the continental margin since about 500 Ma, and the passive continental margin was transformed into an active one (Isozaki et al. 2010). The geohistory of the Japanese Islands has witnessed five distinctive phases: Paleozoic Continental Margin, Paleozoic Island Arc, Paleozoic Accretionary Complex (AC), Mesozoic AC, and Cenozoic Island Arc (Wakita 2013). Accordingly, the major structural events during the geohistory of this region are accretion, paired metamorphism, volcanism, back-arc spreading, and arc–arc collision, which are also vital pieces of the jigsaw puzzle of the geological formation of Asia that we know today (Ibid). Isozaki et al. (2010) divide this geohistory into three main stages of formation: passive continental margin stage (stasis) between 700 Ma and 500 Ma, active continental arc stage between 500 Ma and 20 Ma, and active island arc stage that began since the back-arc opening of the Sea of Japan 20 Ma. (For further information also see Maruyama 1997 and Isozaki et al. 2010.) The geological history of the Japanese Islands is discussed in Chap. 2; here, I provide a brief outline based on the key works on this subject.

Isozaki et al. (2010) provide an excellent description of the processes and mechanisms that contributed to the formation of the archipelago in deep time. The surface crust of the Japanese Islands is comprised of highly deformed sedimentary rocks and metamorphic rocks of predominantly accretionary origin. While the underlying basement is inferred to be of Precambrian origin (Wakita 2013), accretionary complexes that formed during the Mesozoicum form the bulk of the surface crust of Japan (Isozaki et al. 2010). Volcanism has been present in the archipelago since deep time; earliest subduction-related volcanism is identified to have taken place at least 500 Ma (Sakashima et al. 2003). Since the transformation of the passive continental margin into an active one, Japan has witnessed a long history of volcanism and continent–arc collision until 20 Ma. The opening of the Sea of Japan between 20 Ma and 15 Ma formed a distinctive geographical feature as Japan was transformed into an "island arc" that we know today. The most distinctive features of land formation that we see today, i.e., volcanoes and sedimentary formations, notably landforms of the Pacific seaboard, are products of Cenozoic (mainly Neogene and later) volcanism and orogeny. Though these newer formations of the Cenozoic are relatively of lesser importance for understanding the major structural formation phases of the archipelago as pointed out by Isozaki et al. (2010), they are still important to understand the complex crustal processes and volcanism in and around the archipelago (this part of the geohistory of Japan is detailed in Chap. 3). It can certainly be said that Japan has provided an excellent setting for understanding active continental margin formations, accretionary complexes, and continent–arc and arc–arc collision events. Recent studies on accretionary complexes indicate that the growth of the Japanese Islands has not been a uniform process and tectonic erosion played a far bigger role in shaping the geohistory of the islands than previously understood. Suzuki et al. (2010) point out that extensive removal of crusts had taken place in the geological past. This highlights the point that though Japan is generally seen as an island arc with accretionary and volcanically induced "build-up," the real story of its evolution is far more complex with tectonic erosion playing a key part in the rise and fall of landforms, which together make an exceptionally dynamic saga of crustal formation and evolution.

On the longer scale, it is known that supercontinents have assembled and rifted apart at least since 3 Ga, when the supercontinent Ur (Rogers 1996) formed. While considerable uncertainty remains on the formation and extent of the earliest supercontinents, recent geological research has yielded valuable information about supercontinents since the Paleoproterozoic, beginning from the supercontinent Columbia (Santosh et al. 2009). The rifting of the supercontinent Rodinia (which existed approximately between 1.1 Ga and 700 Ma) due to superplume mechanisms (Ibid; Condie 2002) was directly related to the formation of Proto-Japan. Supercontinent formation and breakup, therefore, forms a cycle of earth's crustal evolution, which is complemented by the Wilsonian cycle of opening of mega-oceans (Santosh 2010). It is postulated that 250 million years from now, the Eurasian and North American continental blocks will collide and the Pacific Ocean will close to start continental amalgamation anew, and eventually a new supercontinent, Amasia, will be formed (Maruyama et al. 2007; Santosh et al. 2009, see also Fig. 1.2). While continent–continent collision will consume the Japanese Islands at that point, it has been observed that the nearly 1 billion-year life history of the islands will closely mirror the birth and demise of the Pacific (Isozaki et al. 2010). The birth, evolution, and eventual demise of the Japanese Islands, when seen from this scale of the supercontinent and mega-ocean cycles, is a process that straddles deep time, our contemporaneous geo-biosphere, and the planet of the distant future. While processes and their outcomes stretched over such temporal and spatial scales cannot be fully grasped by the human mind—indeed, life's evolutionary pathways may change so drastically along the unfolding of this process that the our understanding of the geo-biosphere today might have to be replaced with fundamentally different concepts—the geohistory of the Japanese Islands is important in this aspect, as it provides extremely important insights about an old supercontinental fragment in an active dynamic continent–continent, continent–ocean interaction zone. This book aims to provide aspects of this beating pulse of the archipelago by focusing on the different types of "heritage" that are intricately linked with its whole life history.

Geomorphologically, the Japanese Islands represent an interesting case where almost the entire land is simultaneously shaped by prominent uplift, deposition, and denudation processes. While the uplift processes are related to crustal formation and movement, deposition is related to both volcanism (deposition of volcanic material through eruption and redistribution through erosion and transport) and climate (snowfall, rainfall), and swift denudation processes can be observed throughout the islands. What is more, all these processes are intertwined as plate motion uplifts and deforms land and supplies the magma reservoirs, and surface relief is subjected to denudation by agents ranging from fault movement, kinetic displacement of material (slope failures, mountain collapse, dome collapse in case of volcanoes), ground movement of water and ice, precipitation, and wind motion. Yoshikawa (1984) termed the surface Japanese Archipelago as a "tectonically active and

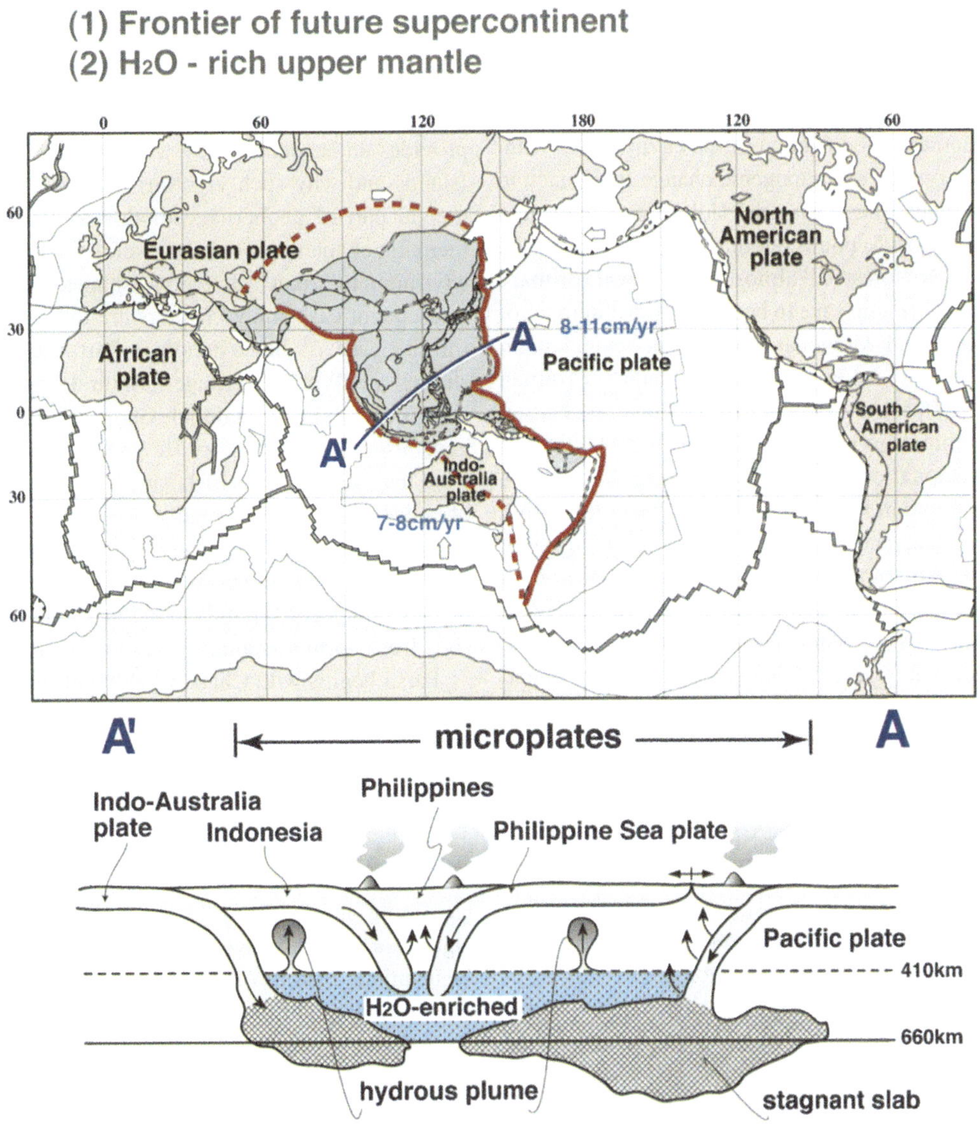

Fig. 1.2 Location of East Asia in a dynamic zone (after Maruyama et al. 2007). The occurrence of double-sided subduction and intense magmatic activity set the stage for opening and closing of oceans and future supercontinent formation. Reprinted from Gondwana Research, 11, Maruyama S, Santosh M, Zhao D, Superplume, supercontinent, and post-perovskite: Mantle dynamics and anti-plate tectonics on the Core–Mantle Boundary, 7–37, Copyright (2007), with permission from Elsevier

intensely denuded" region. The Japanese Alps, which stretch from the northwest to central Honshu Island, form a prime example of Quaternary uplift of mountain blocks due to subduction-related mechanism and intensive denudation.

High precipitation and resultant large stormflow events incise mountainsides to provide a very high sediment yield for Japanese watersheds in general. It is well known that rivers in Japan have typically steep channel profiles compared to continental rivers, which is related to the topographical feature of Japanese mountains that have high relief in spite of small width (Oguchi et al. 2001). The combination of high relief, stormflow events, and occasional tectonic movement results in a high frequency of slope failures and landslides. Two special features of hillslope dynamics in Japanese highlands are: they can occur in densely forested slopes due to the propensity of slopes to move as blocks during high stormflow discharge events; and the relatively rapid reforestation due to wet and temperate climate (Oguchi et al. 2001). High sediment yield and flow in the watersheds transform into widespread alluvial fans in the downstream. Saito (1988) has provided a detailed dataset about the origin and development mechanisms of these fans. Umitsu (1994) provides an account of the thick coastal sedimentation in the Holocene, which means a general lack of incised coastal landforms in the archipelago. Oguchi (1996) further points out that the presence of smooth and incised hillslopes and sediment fill fans in major lowlands describes a process of landscaping borne out of the Pleistocene-Holocene climatic transition.

While the geological and geomorphological processes represent significant natural forces in action, the ecological landscapes of these islands have a very long history of anthropogenic change. It has been observed that due to human habitation and modification of the environment from an early date, pervasive anthropogenic change took place in the East-Asian landscapes during the Holocene (Aikens and Lee 2013). Karan (2005) points out that although 67% of the Japanese Islands is "forested," almost no primeval (primary old growth) forest remains, due to both early development of agriculture (Nasu and Momohara 2016) and several great deforestation waves in early modern history (Totman 1998). Karan (2005) shows that the original vegetation cover in the archipelago is strongly constrained by climatic features and precipitation amount. The four main forest types are Sub-polar (found in Hokkaido and highlands of Honshu, similar to Canadian and North European forests for tree composition), Cool-temperate (found in Western Hokkaido, Honshu; dominant tree type is deciduous), Warm-temperate (evergreen broad-leaf or "laurel" forests of Honshu, Shikoku, and Kyushu, originally very widespread but now rare due to extensive logging), and Sub-tropical (occurring in southern islands and Ogasawara in the east, relatively undisturbed due to distance from major population hubs) (Ibid). Against the backdrop of significant depletion of the old-growth forest cover, some notable primary forest tracts remain standing, such as the relict forests of *Fagus crenata* of Shirakami Mountains that date back to the Last Glacial Maximum (LGM) and the Yakushima Island which has some of the largest and oldest trees of the archipelago.

This book also introduces human connectivity with the natural heritage of the archipelago, which in places resulted in important socio-ecological landscapes such as grasslands and secondary broadleaf forests. Although these socio-ecological "production" landscapes are valued for their capacity to fulfill human needs, they nevertheless have important ecological functions and characteristics. Landscape heterogeneity is a basic feature of these socio-ecological landscapes, many of which have been managed through traditional methods of selective cutting and pruning of vegetation and maintenance of ecological niches. The term "satoyama" (literally meaning "village mountain," indicating managed forests and agricultural landscapes near human habitations) is popularly used for describing such landscapes in Japan (Katoh et al. 2009; Kadoya and Washitani 2011), and increasingly the term "satoumi" (literally "village sea," indicating shallow near coastal marine environments) is used to signify the vital role of the sea for sustenance of island societies (Berque and Matsuda 2013; Henocque 2013). Such landscapes are constantly shaped by the geological and geomorphological forces as well as by human use, and they comprise an important part of the natural heritage of the archipelago.

1.3 A Brief Outline of this Volume

Following this Introduction, Chap. 2 discusses the geohistory of the Japanese Islands in detail. Chapter 3 provides an account of how volcanism shapes the Japanese Islands and why such vigorous natural activity is a fundamental part of geoheritage of this archipelago. These two thematic chapters provide an overall outline of the dynamic environment that actively shapes natural heritage. Subsequent case study chapters are divided into three parts for the ease of understanding. Part I presents four case studies of World Heritage (Natural) properties, Part II describes UNESCO Global Geoparks and challenges of geoconservation, and Part III presents three selected sites from the national parks of Japan. This division is for the ease of understanding; in reality, most WHS overlap with National Park territories and some geoparks also fall under National Park land. The sequence of the chapters in these three sections corresponds to the distribution of the heritage sites from north to south.

Part I begins with Chap. 4, which provides an account of the Shiretoko Peninsula of northeast Hokkaido. A land of active volcanoes, Shiretoko is surrounded by a marine environment that is known to generate the most southerly drift ice conditions in the Northern Hemisphere. Chapter 5 describes the old-growth natural forest of Shirakami Mountains, which is one of the rare old-growth forests remaining in mainland Japan. Chapter 6 introduces the Ogasawara Islands, a remote group of islands known for their unique history of life. Chapter 7 is an analysis of the natural heritage of the Yakushima Island, with an emphasis on the human interaction with the forest landscape.

Part III starts with a discussion of the introduction and the growth of the geopark concept in Japan in Chap. 8. Chapter 9 introduces the San'in-Kaigan Geopark, known for its diverse and beautiful coastal topography and volcanic landforms. The chapter also analyzes an ongoing wildlife restoration scheme in that area. Chapter 10, on Muroto Geopark, provides an account of the area's geological heritage that is defined by spectacular subduction-related landforms. Chapter 11 analyzes the volcanic landscape of Aso, which is one of the largest terrestrial calderas on the planet. Aso is simultaneously characterized by active volcanism and rolling grasslands and is a prime example of a biocultural landscape. Chapter 12 provides an account of the geoheritage of the Unzen (Shimabara) Geopark, a volcanic landscape that has witnessed large volcanic disasters time and again. The chapter describes the most recent volcanic disaster of this area in detail. Chapter 13 provides an analysis of geoconservation challenges currently faced by these geoparks and provides some suggestions for effectively protecting the abiotic diversity of the archipelago.

Part III opens with the account of the Akan National park, one of the earliest national park areas of Japan (Chap. 14). The area is known for its beautiful mountains, winter sport opportunities, scenic volcanic lakes, and a rare colony of marimo algae at Lake Akan. Chapter 15 describes the Oze highland marsh, a place of spectacular natural beauty that helped form the nature conservation movement of Japan. Chapter 16 describes one of the most iconic symbols of the Japanese landscape, the towering stratovolcano of Mount Fuji and its surrounding landscapes.

Finally, Chap. 17 provides a synthesis of all chapters and sums up the key lessons for conservation of natural heritage in this part of the world.

Acknowledgement I thank Shamik Chakraborty and Mahito Watanabe, coeditors of this volume, for providing important insights.

References

Abel ND, Cumming HM, Anderies JM (2006) Collapse and reorganization in socio-ecological systems: questions, some ideas and policy implications. Ecol Soc 11(1):17

Aikens CM, Lee GA (2013) Postglacial inception and growth of anthropogenic landscapes in China, Korea, Japan, and the Russian far East. Anthropocene 4:46–56

Autin WJ, Holbrook JM (2012) Is Anthropocene an issue of stratigraphy or pop culture? GSA Today 22:7. doi:10.1130/G153GW.1.

BBC News (2017) Crime threatens nearly half of World's natural heritage sites. http://www.bbc.com/news/science-environment-39590117. Accessed 13 May 2017

Berque J, Matsuda O (2013) Coastal biodiversity management in Japanese Satoumi. Mar Policy 39:191–200

Castree N (2005) Nature. Routledge, London

Clark N (2011) Inhuman nature: sociable life on a dynamic planet. Sage, London

Condie KC (2002) Breakup of a Paleopropterozoic supercontinent. Gondw Res 4:41–43

Crutzen PJ, Stoermer EF (2000) The Anthropocene. Glob Chang Newslett 41:17–18

Gray M (2013) Geodiversity: valuing and conserving abiotic nature. Wiley, Chichester

Gregory KJ, Lewin J (2014) The basics of geomorphology. Sage, London

Gunderson LH, Holling CS (2001) Panarchy: understanding transformations in human and natural systems. Island Press, Washington, DC

Gunderson LH, Allen CR, Holling CS (2010) Foundations of ecological resilience. Island Press, Washington, DC

Henocque Y (2013) Enhancing social capital for coastal development: is Satoumi the answer? Estuar Coast Shelf Sci 116:66–73

Isozaki Y, Aoki K, Nakama T, Yanai S (2010) New insight into a subduction-related Orogen: a reappraisal of the geotectonic framework and evolution of the Japanese Islands. Gondw Res 18:82–105

Kadoya T, Washitani I (2011) The Satoyama Index: a biodiversity indicator for agricultural landscapes. Agric Ecosyst Environ 140 (1):20–26

Karan P (2005) Japan in the 21st century: environment, economy and society. University Press of Kentucky, Lexington, KY

Katoh K, Sakai S, Takahashi T (2009) Factors maintaining species diversity in Satoyama: a traditional agricultural landscape of Japan. Biol Conserv 142(9):1930–1936

Malm A, Hornborg A (2014) The geology of mankind? A critique of the Anthropocene narrative. Anthropocene Rev 1(1):62–69

Maruyama S (1997) Pacific type orogeny revisited: Miyashiro type orogeny proposed. Island Arc 6:91–120

Maruyama S, Isozaki Y, Kimura G, Terabayashi S (1997) Paleogeographic maps of Japanese Islands: plate tectonic synthesis from 750 Ma to the present. Island Arc 6:121–142

Maruyama S, Santosh M, Zhao DP (2007) Superplume, supercontinent and post-perovskite: mantle dynamics and anti-plate tectonics on the core-mantle boundary. Gondw Res 11:7–37

Nasu H, Momohara A (2016) The beginnings of rice and millet agriculture in prehistoric Japan. Quat Int 397:504–512

Oguchi T (1996) Factors affecting the magnitude of post-glacial hillslope incision in Japanese Mountains. Catena 26:171–186

Oguchi T, Saito K, Kadomura H, Grossman M (2001) Fluvial geomorphology and paleohydrology in Japan. Geomorphology 39:3–19

Payne AJ, Vieli A, Shepherd AP, Wingham DJ, Rignot E (2004) Recent dramatic thinning of largest West Antarctic Ice Stream triggered by oceans. Geophys Res Lett. http://onlinelibrary.wiley.com/doi/10.1029/2004GL021284/full. Accessed 8 Aug 2016

Rignot E, Jacobs SS (2002) Rapid bottom melting widespread near Antarctic ice sheet grounding lines. Science 296(5575):2020–2023

Rogers JJW (1996) A history of continents in the past three billion years. J Geol 104(1):91–107

Saito K (1988) Nihon no Senjochi (Alluvial fans of Japan). Kokon Shoin, Tokyo

Sakashima T, Terada K, Takeshita T, Sano Y (2003) Large scale displacement along the median Tectonic Line, Japan: evidence from the SHRIMP Zircon U-Pb dating of granites and Gneisses from the South Kitakami and Paleo Ryoke Belts. J Asian Earth Sci 21:1019–1039

Santosh M (2010) Supercontinent tectonics and biogeochemical cycle: a matter of 'life and death'. Geosci Front 1:21–30

Santosh M, Maruyama S, Yamamoto S (2009) The making and breaking of supercontinents: some speculations based on superplumes, superdownwelling and the role of Tetctosphere. Gondw Res 15:324–341

Suzuki K, Maruyama S, Yamamoto S, Omori S (2010) Nihon Retto Wa Seicho Shitanoka?: 5tsu no Nihon ga Umare 4tsu no Nihon ga Shizumikomi Shoushitsu Shita (Have the Japanese Islands grown?: Five Japans were born and four Japans Subducted in to the Mantle). Chigaku Zasshi (J Geogr) 119(6):1196–2010 (In Japanese)

Totman C (1998) The Green Archipelago: forestry in preindustrial Japan. Ohio University Press, Athens, OH

Umitsu M (1994) Chuseki Teichi no Kokankyogaku (Late quaternary environment and landform evolution of riverine coastal lowlands). Kokon Shoin, Tokyo

UNESCO (2006) UNESCO World Heritage Centre's natural heritage strategy. http://whc.unesco.org/en/naturalheritagestrategy/. Accessed 8 Aug 2016

Wakita K (2013) Geology and tectonics of Japanese Islands: a review-the key to understanding the Geology of Asia. J Asian Earth Sci 72:75–87

Waters CW et al (2016) The Anthropocene is functionally and stratigraphically distinct from the Holocene. Science 351 (6269). http://science.sciencemag.org/content/351/6269/aad2622. Accessed 12 Jan 2017

Wohl E (2013) Wilderness is dead: Whither critical zone studies and geomorphology in the Anthropocene? Anthropocene 2:4–15

Yoshikawa T (1984) Shitujun Hendotai no Chikeigaku (Geomorphology of tectonically active and intensely denuded regions). Chirigaku Hyoron (Geogr Rev Jpn) 57A:691–702 (In Japanese)

Geology of the Japanese Islands: An Outline

2

Koji Wakita

Abstract

The geological history of the Japanese Islands began from the breakup of the Supercontinent Rodinia about 750 Ma. After a passive margin setting of about 250 million years, the Paleo-Pacific Ocean started to subduct beneath the Paleo-Asian continent about 500 Ma. The tectonic setting of this archipelago is an active convergent margin, where an oceanic plate has been continually subducting for over 500 million years. Plate subduction formed accretionary complexes based on sediment supply from mountains developed in the active continental margins. The accretionary complex is characterized by "ocean plate stratigraphy" which is basalt, limestone, chert, siliceous shale, and terrigenous turbidite in the ascending order. Non-metamorphosed accretionary complexes range from Carboniferous to the present in age. The accretionary complexes are metamorphosed into metamorphic rocks of low-temperature and high-pressure type and of high-temperature and low-pressure type. Extensive igneous activities occurred during Cretaceous to Paleogene, and arc volcanism has been very active throughout Cenozoic. The major tectonic events in Cenozoic are back-arc spreading and arc–arc collision. These events formed the Sea of Japan as a back-arc basin and a complicated island arc system around Japan. Plate subduction also causes frequent earthquakes and volcanic activities in the Japanese Islands. Although Japanese Islands suffer various geological hazards, these geological processes are fundamental for the formation of land for us to live on and for beautiful sceneries like Mt. Fuji.

Keywords

Japanese Islands • Active convergent margin • Supercontinent breakup • Ocean plate stratigraphy • Accretionary complexes • Arc–arc collision

2.1 Introduction

The geological history of the Japanese Islands is recorded in various rocks and formations distributed on the surface of present Japanese Islands. The age of rocks and formations ranges from 500 Ma (early Late Proterozoic) to the present (Fig. 2.1). The geology of Japan developed at the continental margin, along the eastern margin of the Asian continent. Among its whole history, its tectonic setting has been that of an active convergent margin, where an oceanic plate is continually subducting for over 500 million years. There is no other place in the world that has been in active geological setting for such a long time. As this process has unfolded across geological time, the geology of the archipelago has been characterized by sediment accretion, faulting and folding, arc volcanism and plutonism, and paired metamorphism along the convergent margin (Wakita 2013). The basement of the Japanese Islands is mainly composed of accretionary

K. Wakita (✉)
The Graduate School of Sciences and Technology for Innovation, Yamaguchi University, Yamaguchi, Japan
e-mail: k-wakita@yamaguchi-u.ac.jp

A. Chakraborty et al. (eds.), *Natural Heritage of Japan*, Geoheritage, Geoparks and Geotourism,
DOI 10.1007/978-3-319-61896-8_2

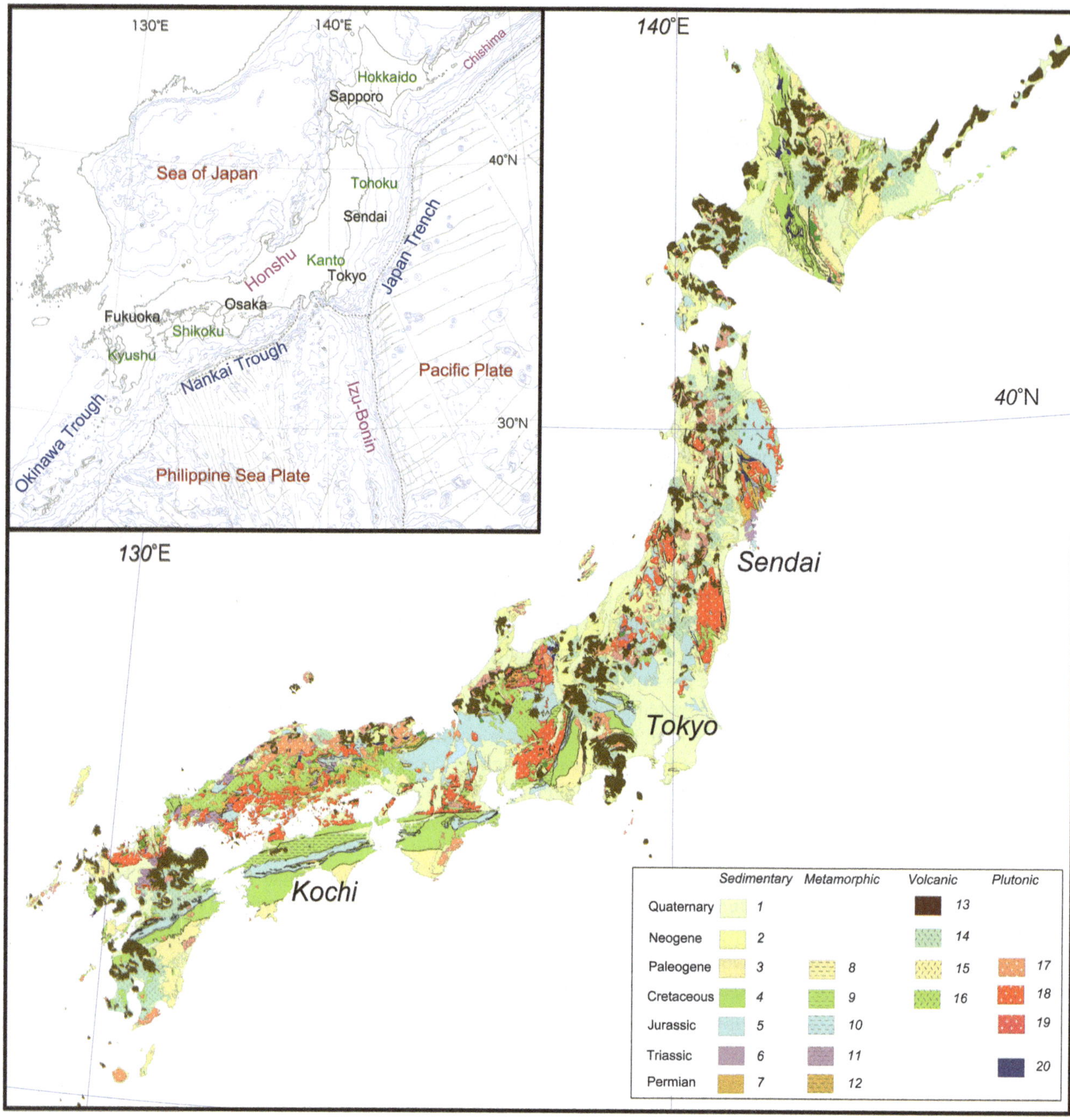

Fig. 2.1 Geological map of Japan (Wakita 2013). Reprinted from Journal of Asian Earth Sciences 72, Wakita K Geology and tectonics of Japanese Islands: A review – The key to understanding the geology of Asia, 75–87, Copyright (2013), with permission from Elsevier

complexes of various ages (Fig. 2.2). Arc volcanism and related plutonism are also caused by plate subduction. Mesozoic and Cenozoic volcanic and plutonic rocks are distributed on the Japanese Islands. Cenozoic tectonism is characterized by back-arc spreading that formed the Sea of Japan and arc–arc collision. The Miocene back-arc opening formed volcanic rocks (Green Tuff) and hydrothermal ore deposits (Kuroko) in the extensional setting. Currently, arc–arc collision is ongoing between the Honshu and Izu-Bonin Arcs.

2.2 Super-continent Breakup and Beginning of Subduction

The first step of the geological history of Japan is the breakup of the super-continent "Rodinia" about 750 Ma (Fig. 2.3). Upon breakup, the super-continental landmass fragmented into smaller continents, including Laurentia (the present North American region) and the Paleo-Asian continent (mainly the South China region). The Paleo-

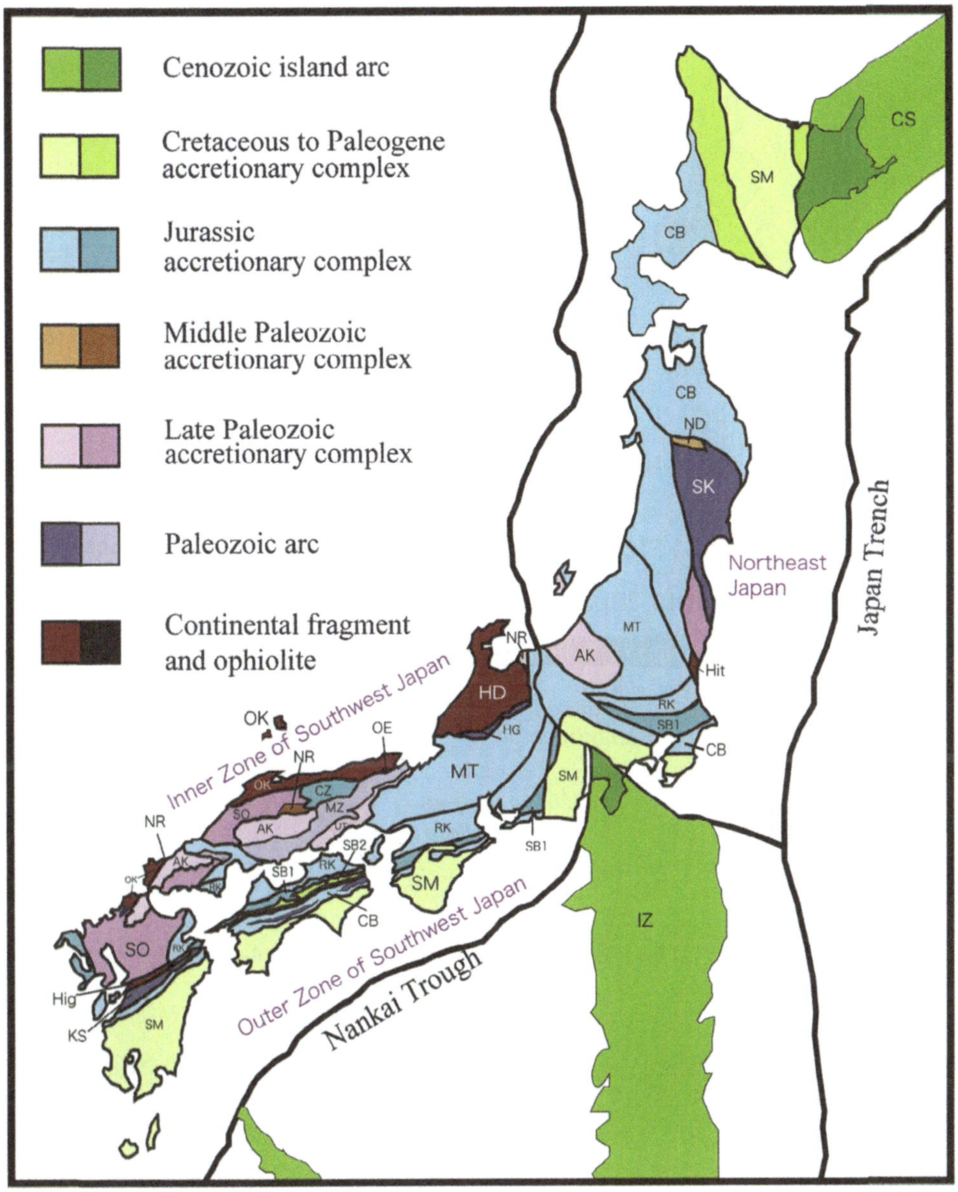

Fig. 2.2 Distribution of accretionary complexes of the Japanese Islands. *AK* Akiyoshi Belt, *CB* Chichibu Belt, *CS* Chishima Arc, *CZ* Chizu Belt, *HD* Hida Belt, *HG* Hida Gaien Belt, *HH* Higo-Hitachi Belt, *HS* Honshu Arc, *IZ* Izu Arc, *KS* Kurosegawa Belt, *MT* Mino-Tamba Belt, *MZ* Maizuru Belt, *NR* Nagato-Renge Belt, *ND*: Nedamo Belt, *OE* Oeyama Ophiolite, *OK* Oki Belt, *RK* Ryoke Belt, *SB* Sanbagawa Belt (SB1: mid-Cretaceous; SB2: Late Cretaceous), *SK* South Kitakami Belt, *SM* Shimanto Belt, *SO* Suo Belt, *UN* Unazuki Belt, *UT* Ultra-Tamba Belt (after Wakita 2013). Reprinted from Journal of Asian Earth Sciences 72, Wakita K Geology and tectonics of Japanese Islands: A review – The key to understanding the geology of Asia, 75–87, Copyright (2013), with permission from Elsevier

Pacific Ocean formed between Laurentia and the Asian continent about 750 Ma (Maruyama et al. 1997; Isozaki et al. 2010). At first, the boundary between the Asian continent and the Paleo-Pacific Ocean was passive margin like the present Atlantic Ocean margin.

The Paleo-Pacific Ocean started to subduct beneath the Paleo-Asian continent about 500 Ma (Fig. 2.2). Consequently, the eastern margin of the Asian continent became an active convergent margin where various rocks and sediments formed. The Oeyama Ophiolite complex is one of the remnants of the oceanic plate at that time and is composed mainly of peridotite, dunite, harzburgite, and hornblendite with metagabbro and diabase. The geological record in Japan is deficient in the Early to Middle Paleozoic, but these formations are dominated by low-temperature–high-pressure metamorphic rocks of about 400 Ma, very old granite of 440 Ma, and metamorphic rocks of about 500 Ma. They exist as fault bounded fragments in various tectonic belts such as the Hida-Gaien tectonic belt, Kurosegawa tectonic belt, Hitachi belt, and the South Kitakami belt. These tectonic fragments were derived from tectonic erosion from the ancient Asian continental margin

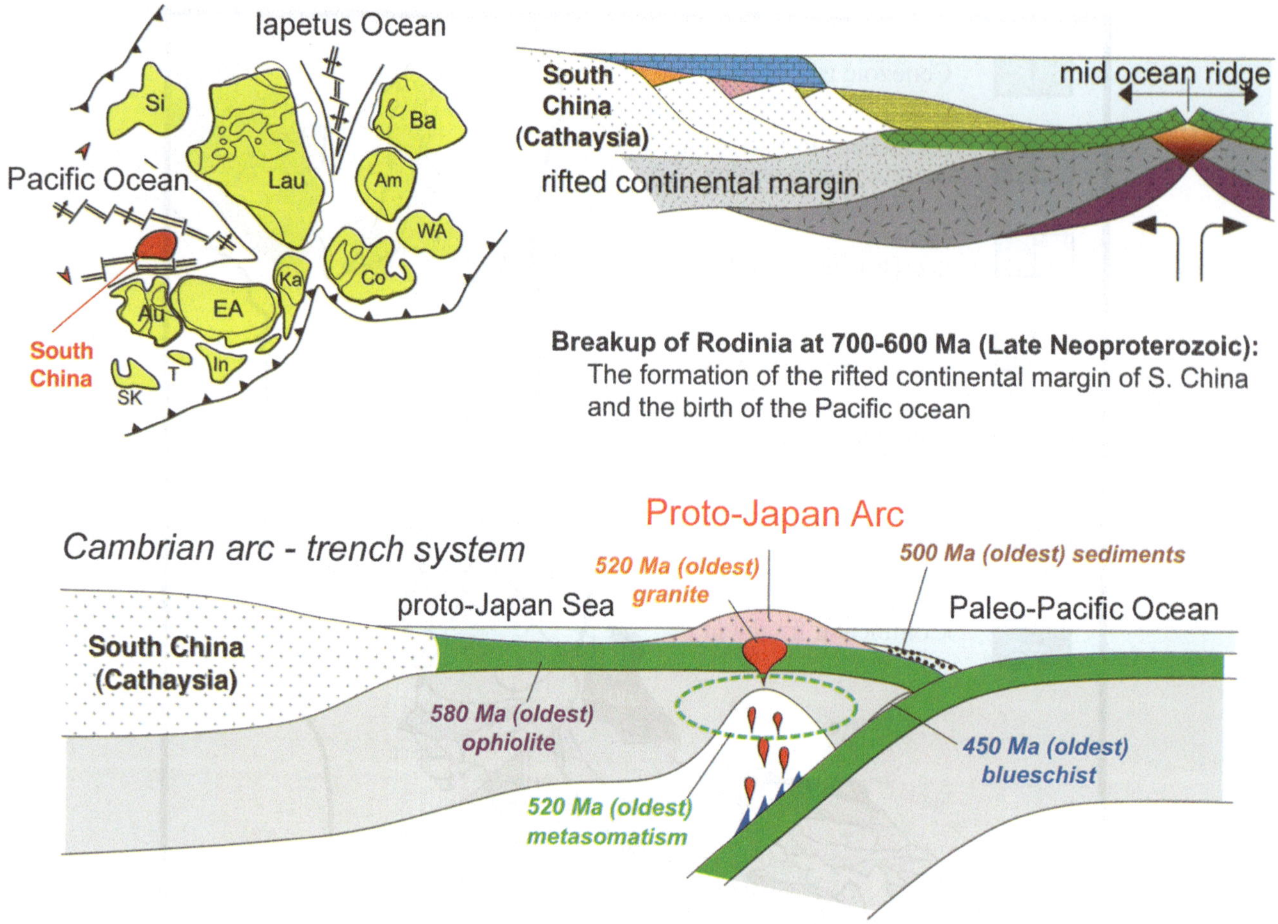

Fig. 2.3 Breakup of Rodinia (about 750 Ma) and the start of subduction (about 500 Ma). Modified from Fig. 6 of Isozaki et al. (2011)

(Isozaki et al. 2010); the tectonic fragments were exposed within serpentine melanges of tectonic belts.

2.3 Accretionary Complexes

As the foregoing discussion shows, the geohistory of the Japanese Islands has been dominated by accretion and tectonic erosion. The major parts of the basement are composed of the accretionary complexes of late Paleozoic to Cenozoic (Isozaki et al. 2010; Wakita 2013). Mountain building and arc volcanism along the Paleo-Asian continental margin supplied large amounts of terrigenous sediments to the trench where the ocean plate had been subducting. These sediments are accreted to the continental side to form the accretionary complexes, which form the origin of Japanese basement (Fig. 2.4). The accretionary complexes are composed of "ocean plate stratigraphy," a geological term that describes the stratigraphic succession composed of the rocks and formations of the oceanic plate and the terrigenous sediment deposited at trench sites (Isozaki et al. 1990; Matsuda and Isozaki 1991; Wakita and Metcalfe 2005). They are composed mainly of basalt, limestone, chert, siliceous shale, and terrigenous turbidite in this ascending order. Basalt was detached from the upper part of the volcanic islands, limestone formed as calcareous reefs on the volcanic islands, and chert and siliceous shale are pelagic and hemipelagic sediments, respectively. Chert is mainly composed of siliceous skeletons of oceanic planktons. Siliceous shale is a mixture of such siliceous skeletons and terrigenous mud. Terrigenous turbidite was derived from the land area of the continental margin and deposited as trench-fill sediments. These rock formations of the "ocean plate stratigraphy" were scraped off from the subducting ocean plate by decollement near the trench. The repetition of off-scraping and under-plating of the "ocean plate stratigraphy" eventually formed tectonically stacked "ocean plate stratigraphy" or its chaotic mixtures such as melanges that together construct accretionary complexes. These are the major basement components of the Japanese Islands. A brief synopsis of this accretionary complex formation and evolution is given below.

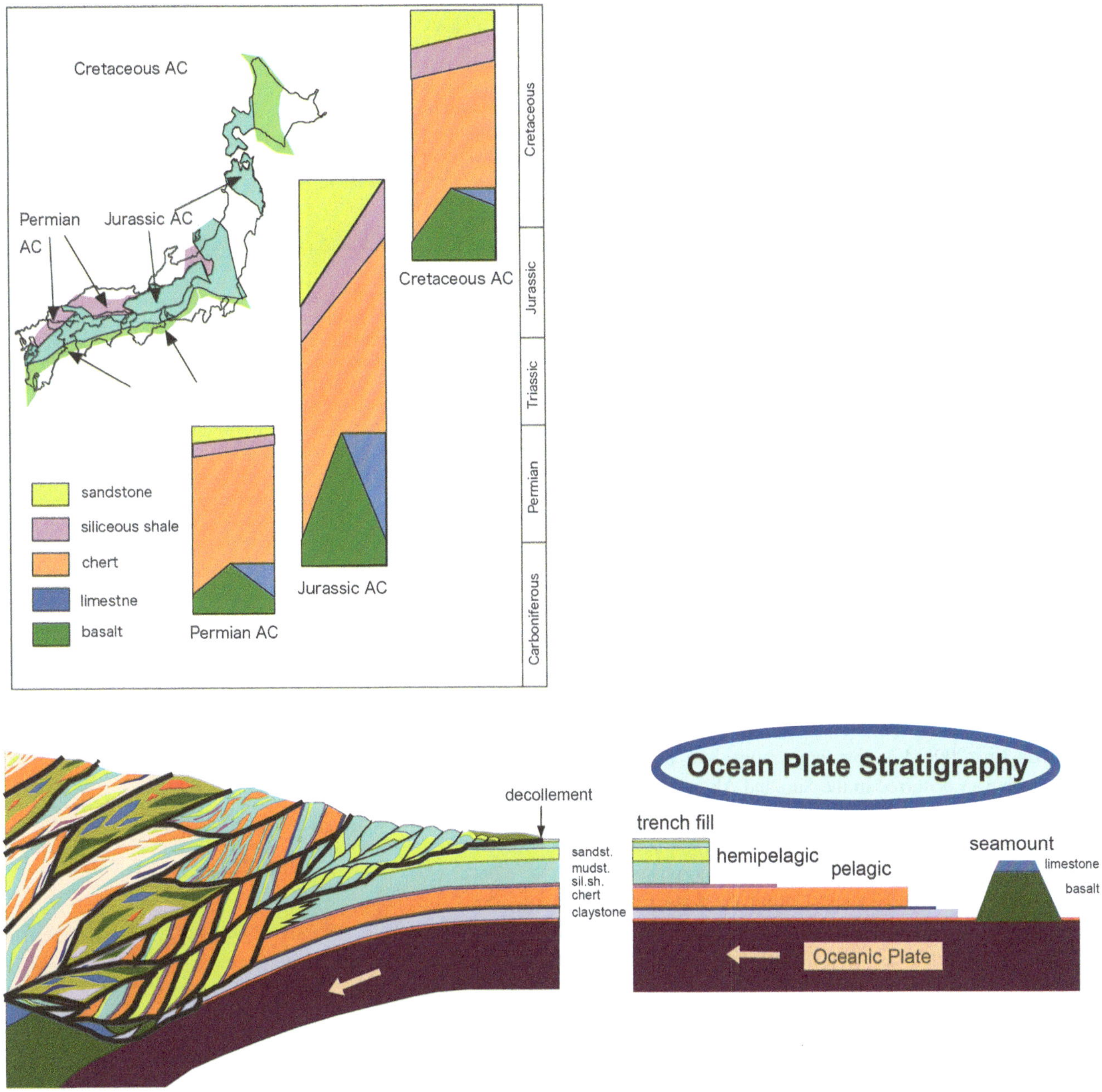

Fig. 2.4 Distribution of three major accretionary complexes (*Left*) and the formation process of accretionary complex from "Ocean Plate Stratigraphy" (Wakita 2012, 2013). Reprinted from Journal of Asian Earth Sciences 72, Wakita K Geology and tectonics of Japanese Islands: A review – The key to understanding the geology of Asia, 75–87, Copyright (2013), with permission from Elsevier

The oldest non-metamorphosed accretionary complex is of Carboniferous age in the Nedamo Belt. The protolith of the low-temperature high-pressure metamorphic rocks of the Nagato-Renge Belt was incorporated into the accretionary complexes of Middle to Late Paleozoic. Middle to Late Permian accretionary complexes are distributed in the Akiyoshi and Ultra-Tamba belts. The components of these accretionary wedges are Carboniferous to Middle Permian limestone associated with basalt, Permian chert including radiolarian remains and sponge spicules, felsic tuff, sandstone, mudstone, and conglomerate. The Maizuru belt adjacent to the Akiyoshi and Ultra Tamba belts was a remnant of a Late Paleozoic island arc. It is composed of mudstone, metabasalt, amphibolite, metagabbro, metagranite, and ultramafic rocks (Ishiwatari and Tsujimori 2003).

Late Triassic to Early Cretaceous accretionary complexes are characterized by the dominance of radiolarian cherts. "Ocean plate stratigraphy" was separated from the underlying ocean floor by a decollement in the super-anoxic claystone at the Permian-Triassic (P-T) Boundary (Isozaki 1993). Therefore, sediments younger than the P-T boundary claystone were accreted and formed the accretionary complexes, which also contain radiolarian chert and hemipelagic siliceous shale as well as trench turbidites. On the other hand, the lower parts of "ocean plate stratigraphy" were underplated and formed melanges dominated by basalt and limestone. The Shimanto belt is composed of a Cretaceous to Paleogene accretionary complex that is dominated by turbidite and contains comparatively less chert and limestone. Neogene to Quaternary accretionary complexes are mainly exposed offshore along the Nankai Trough, except for the southern margin of Shikoku and the southern margin of Kanto area.

2.4 Metamorphic and Igneous Events

Metamorphic events in Japan are divided into low-temperature and high-pressure and high-temperature and low-pressure types. Medium-pressure metamorphism is recognized only in limited areas such as in the Unazuki, Higo, and Hitachi belts, where ocean plates had subducted beneath Japan from about 500 Ma (Fig. 2.5). Low-temperature and high-pressure metamorphism has occurred throughout the Phanerozoic. Paleozoic metamorphism is recorded in the Hayachine complex (421–484 Ma), the Nishidohira metamorphic rocks (511 Ma) in the South Kitakami Belt, the Terano metamorphic rocks (400 Ma) in the Kurosegawa Belt, and the Nagato-Renge metamorphic rocks (about 320 Ma). During the Triassic (200 Ma), low-temperature and high-pressure metamorphism occurred in the Suo and Akiyoshi belts. During the mid-Cretaceous (110–120 Ma), the earliest Cretaceous accretionary complex of the Sanbagawa belt was subjected to low-temperature and high-pressure metamorphism of the greenschist to eclogite facies. During the Late Cretaceous (60–70 Ma), the mid-Cretaceous accretionary complex of the Shimanto belt was subjected to low-temperature and high-pressure metamorphism.

Low-temperature and high-pressure metamorphism is characterized by the Sanbagawa belt, whereas high-temperature and low-pressure metamorphism occurred in the Hida and Ryoke belts, caused by granitic intrusion. Similar metamorphism occurred in the Unazuki Belt where medium-pressure type metamorphism was overprinted in the Permian. During the mid-Cretaceous (100 Ma), high-temperature and low-pressure metamorphism (termed Ryoke metamorphism) affected some parts of the Jurassic and Permian accretionary complexes. The major protolith of these metamorphic rocks are "ocean plate stratigraphy." Seamount basalt became mafic schist or gneiss, and seamount limestone turned into calcareous schist or gneiss. Psammitic, pelitic, and siliceous schist were derived, respectively, from sandstone, mudstone, and chert of the "ocean plate stratigraphy" accreted in the accretionary complexes.

In contrast, igneous activity mainly occurred during the Paleozoic, Mesozoic, and Cenozoic (Fig. 2.5). Early Paleozoic igneous activity is recorded as Cambrian granite

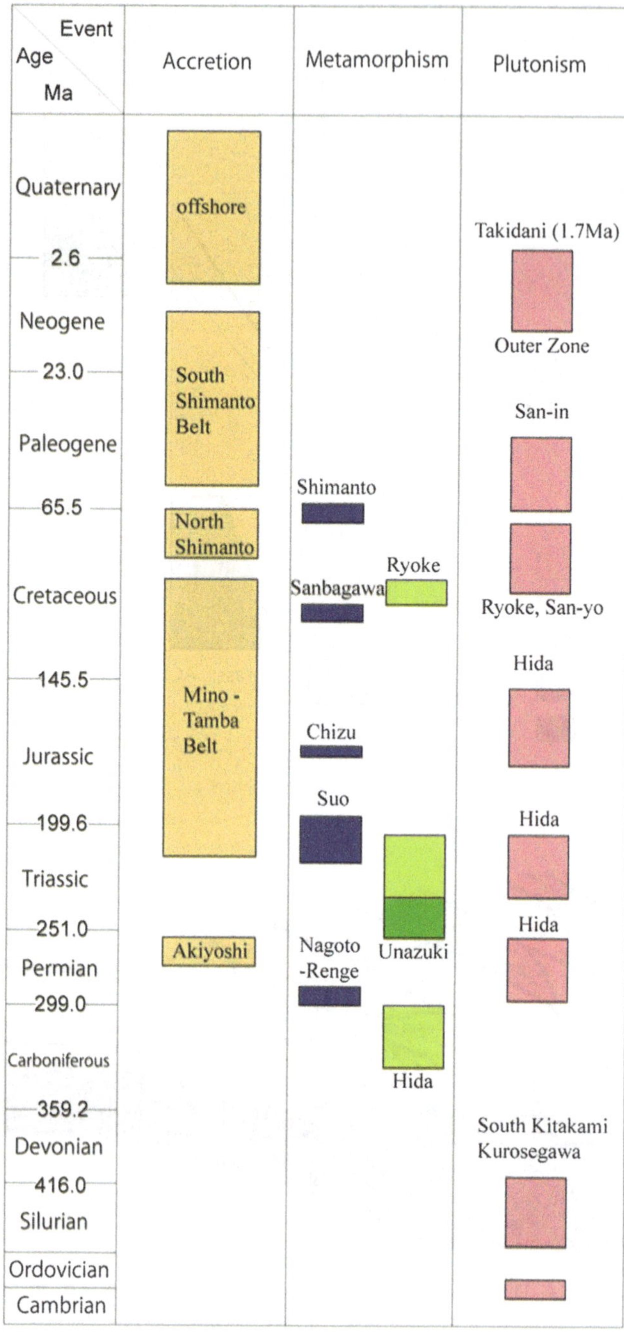

Fig. 2.5 Accretion, metamorphism, and igneous activities through the geological history of the Japanese Islands (Wakita 2013). Reprinted from Journal of Asian Earth Sciences 72, Wakita K Geology and tectonics of Japanese Islands: A review – The key to understanding the geology of Asia, 75–87, Copyright (2013), with permission from Elsevier

(520 Ma) and the Hikami Granite (412 Ma) of the South Kitakami Belt. Late Paleozoic igneous activity is identified in the Hida metagranites of 330–300 Ma and 270–250 Ma in the Hida Belt. Triassic to Jurassic (220–180 Ma) igneous activity is overprinted on the Hida Belt. Lower Cretaceous igneous activity formed the Kwanmon and Sasayama Groups. During the Cretaceous and Paleogene, extensive igneous activity occurred throughout the archipelago

(Fig. 2.6). Major plutons from this phase are the Ryoke (100–70 Ma), Hiroshima (San-yo) (110–70 Ma), San-in (68–37 Ma), Abukuma (110–90 Ma), and Kitakami Granites (120–110 Ma). These plutons are mainly coarse- to medium-grained biotite granite, granodiorite, and tonalite, associated with quartz diorite and pegmatite.

2.5 Back-Arc Spreading, Formation, and Collision of Island Arcs

The geological entities of the Japanese Islands mentioned above are formed along the eastern margin of Asian continents at the continental margin. Another major feature in the geologic formations is the back-arc spreading of an oceanic plate. During this process, East Japan was rotated counterclockwise, while Southwest Japan was rotated clockwise. The rotation began 20 Ma and stopped 15 Ma. The Sea of Japan formed due to this back-arc spreading process, and the Japanese Islands became an island arc. At the same time, the Shikoku Basin and Chishima (Kuril) Basin started to open.

During the back-arc opening, extensive volcanic activities occurred. These activities resulted in intermediate to felsic volcanic rocks and shallow marine to non-marine sedimentary formations. These rocks and formations are commonly called "Green Tuff." The "Green Tuff" includes the famous "Kuroko" ore deposits which were caused by hydrothermal activities at the back-arc setting.

The main components of Cenozoic Japan are four major island arcs, i.e., Honshu Arc, Ryukyu Arc, Izu-Bonin Arc,

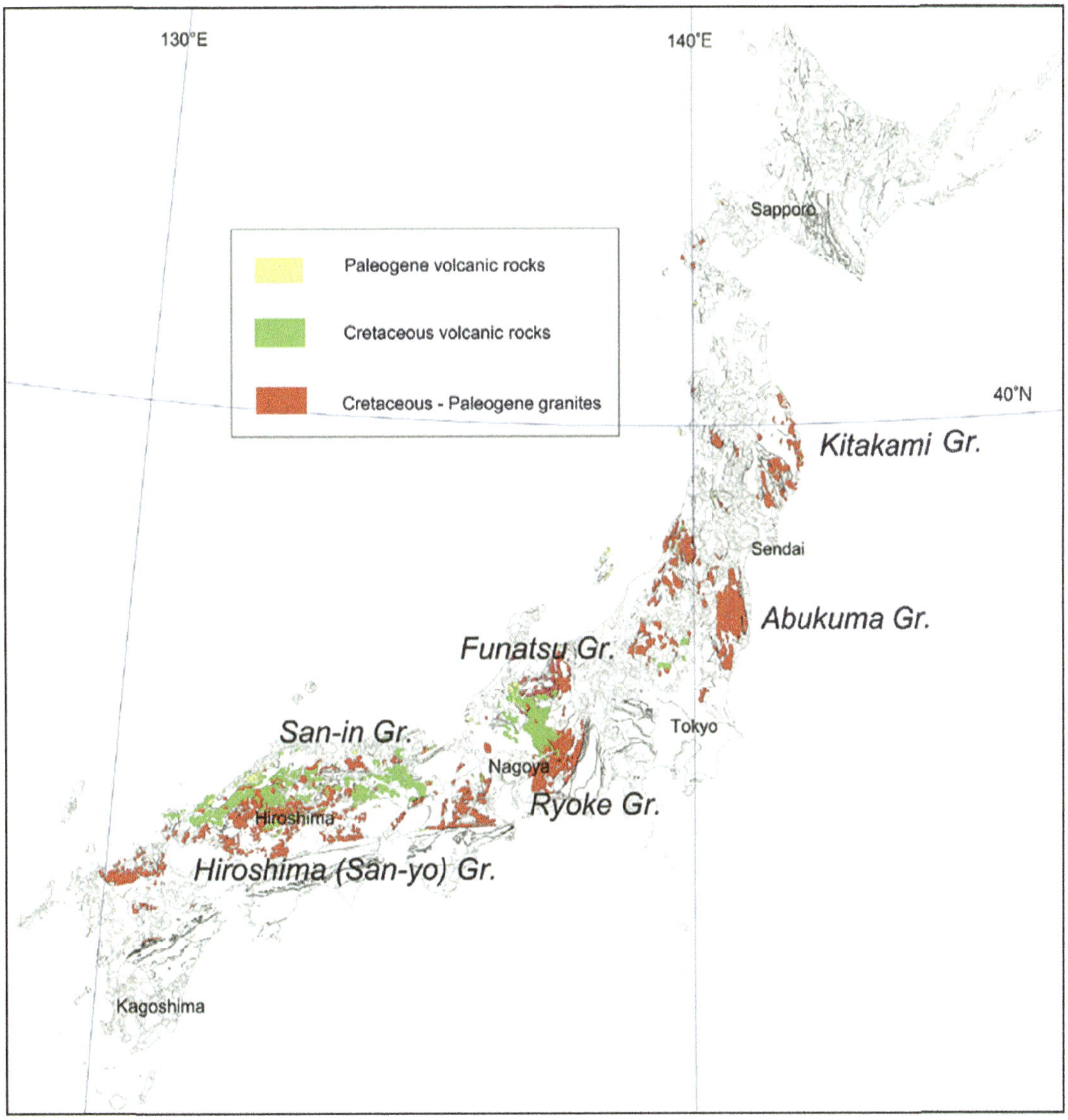

Fig. 2.6 Distribution of Cretaceous-Paleogene igneous rocks of the Japanese Islands. Reprinted from Journal of Asian Earth Sciences 72, Wakita K Geology and tectonics of Japanese Islands: A review – The key to understanding the geology of Asia, 75–87, Copyright (2013), with permission from Elsevier

and Kuril Arc. The Honshu Arc is composed of Paleozoic to Cenozoic rocks accreted by ocean plate subduction along the Asian continental margin and is divided into the Southwest Honshu (or Japan) Arc and the Northeast Honshu (or Japan) Arc by the Fossa Magna in central Japan. The Pacific Plate is subducting along the Japan Trench beneath the northeastern part of the Honshu Arc. The subduction of the cold and old oceanic plate results in tectonic erosion along the Japan Trench. Huge amounts of rock previously accreted were removed by tectonic erosion (Suzuki et al. 2010). On the other hand, the young Philippine Sea Plate is being subducted at Nankai Trough in Southwest Japan and forms an accretionary wedge along the subduction zone.

The Ryuku Arc is underlain by Paleozoic and Mesozoic accretionary complexes belonging to the Nagato-Renge, Chichibu, and Shimanto belts. The Okinawa Trough, a back-arc basin, lies along the northwestern side of the Ryukyu Arc. This arc extends southwestward as far as Taiwan, where it connects with the Philippine Arc. The Izu-Bonin volcanic arc extends for about 1200 km and has a mean width of about 400 km. The Chishima (Kuril) Arc, formed by the subduction of the Pacific Plate beneath the Okhotsk Plate, extends from the Eastern part of Hokkaido to the Aleutian Islands via the Kuril Islands. The basement of the arc is formed of Cretaceous igneous and sedimentary rocks, some of which are exposed in eastern Hokkaido. Cenozoic volcanic and sedimentary rocks cover the Cretaceous basement extensively.

During the Cenozoic era, arc–arc collision occurred between the Honshu and Izu-Bonin (Izu-Ogasawara) and the Honshu and Chishima (Kuril) arcs. The Izu-Bonin Arc collided with the Honshu Arc during the Middle Miocene (15 Ma) (Sugimura 1972). The Izu-Bonin Arc is pushing the Honshu arc northward, together with the migration of the Philippine Sea Plate (Tamura 2011). The collision between the Honshu Arc and the Chishima Arc was another major arc–arc collision (Kimura 1986; Komatsu et al. 1989) that was caused by the westward movement of the Okhotsk Plate against the Eurasia Plate. This arc–arc collision was the main cause of the uplift of the Hidaka Mountains in central Hokkaido.

2.6 Conclusion: The Japanese Islands of Today

At the Quaternary, the formation of this archipelago still continues. While we live on Quaternary alluvial plains formed by the activities of major rivers of Japan, the Pacific Plate continues to subduct beneath East Japan, causing intermittent volcanic activities at the backbone of East Japan. The eastward movement of East Japan and the westward movement of the Pacific Plate cause compression and major earthquakes, the most significant of which was the Magnitude 9.0 East Japan Earthquake of 2011. This megaquake was triggered by the subduction of the Pacific Plate beneath the Honshu Arc along the Japan Trench. The reverse fault movement between the Honshu Arc and the Pacific Plate has caused mega-tsunamis over geological time. Even as we write about the active geological history of this archipelago, a major earthquake is expected along the Nankai Trough where the Philippine Sea Plate is subducting beneath West Japan. The subducting oceanic plate has caused not only volcanism and earthquakes but also sediment accretion and tectonic erosion in the past. In West Japan, the Philippines Sea Plate is subducting along the Nankai Trough where sediment accretion forms new continental crust in the forearc region. In East Japan, the Pacific Plate is subducting along the Japan Trench and the Izu-Ogasawara Trench, where tectonic erosion occurs to remove Japanese continental margin in the offshore area.

These geological activities over 750 million years have provided us the fertile and rich undulating land and beautiful scenery of Japan of today. The Tokyo metropolitan area is underlain by a Quaternary alluvial plain, to which huge amounts of sediments have been provided by surrounding volcanoes. Mount Fuji, the most famous volcano in Japan, is the result of arc volcanism caused by plate subduction. Japanese geoparks are in turn mostly related to the activities of plate subduction such as volcanoes, accretionary processes, high-pressure metamorphism, and back-arc spreading. Although Japanese Islands suffer various geological hazards, volcanic eruptions, earthquakes, and landslides, these geological processes also supply sediments to the ocean to form new land for us to live on. Without the subduction of oceanic plates causing volcanic eruptions and earthquakes, land formation would not have been possible in this area.

As described in this brief sketch, the geological history of the Japanese Islands is a complex one and one that is still unfolding in an active tectonic setting. This has resulted in natural hazards from time to time but also provides ample source of geothermal energy in the forms of volcanoes and hot springs. The geological landscape of the Japanese Islands is underpinned by the restless activity of our planet, which makes this a singularly interesting place for earth scientists.

References

Ishiwatari A, Tsujimori T (2003) Paleozoic ophiolites and blueschists in Japan and Russian primorye in the tectonic framework of East Asia. Island Arc 12:190–206

Isozaki Y (1993) Superanoxia across the P/T Boundary and Mass Extinction. EOS 74:330

Isozaki Y, Aoki K, Nakama T, Yanai S (2010) New insight into a subduction-related orogen: a reappraisal of the geotectonic framework and evolution of the Japanese Islands. Gondwana Res 18:82–105

Isozaki Y, Maruyama S, Fukuoka F (1990) Accreted oceanic material of Japan. Tectonophysics 181:179–205

Isozaki Y, Maruyama S, Nakama T, Yamamoto S, Yanai S (2011) Growth and shrinkage of an active continental margin: updated Geotectonic history of the Japanese Islands. Chigaku Zasshi (J Geogr) 120:65–99 (In Japanese with English abstract)

Kimura G (1986) Oblique subduction and collision: Forearc tectonics of the Kuril Arc. Geology 14:404–407

Komatsu M, Osanai Y, Toyoshima T, Miyashita S (1989) Evolution of the Hidaka Metamorphic Belt, Northern Japan. Geol Soc Lond Spec Publ 43:487–493

Maruyama S, Isozaki Y, Kimura G, Terabayashi M (1997) Paleogeographic maps of the Japanese Islands: Plate tectonic synthesis from 750 ma to the present. Island Arc 6:121–142

Matsuda T, Isozaki Y (1991) Well-documented travel history of Mesozoic Pelagic Chert in Japan: from remote ocean to subduction zone. Tectonics 10:475–499

Sugimura A (1972) Nihon Fukin ni Okeru Pureto no Kyokai (Plate boundaries in Japan and its adjacent areas). Kagaku (Science) 42:192–202 (In Japanese)

Suzuki K, Maruyama S, Yamamoto S, Omori S (2010). Nihon Retto Wa Seicho Shitanoka?: 5tsu no Nihon ga Umare 4tsu no Nihon ga Shizumikomi Shoushitsu Shita (Have the Japanese Islands grown?: five Japans were born and four Japans subducted in to the mantle). Chigaku Zasshi (J Geogr) 119(6):1196–2010 (In Japanese)

Tamura Y (2011) Formation of continental crust at the Izu-Honshu Collision Zone. Chigaku Zasshi (J Geogr) 120:567–584 (In Japanese with English abstract)

Wakita K (2012) Mappable features of Mélanges derived from ocean plate stratigraphy in the jurassic accretionary complexes of Mino and Chichibu Terranes in Southwest Japan. Tectonophysics 568–569:74–85

Wakita K (2013) Geology and tectonics of Japanese Islands: a review: the key to understanding the geology of Asia. J Asia Earth Sci 72:75–87

Wakita K, Metcalfe I (2005) Ocean plate stratigraphy in East and Southeast Asia. J Asia Earth Sci 24:670–702

Volcanic Archipelago: Volcanism as a Geoheritage Characteristic of Japan

3

Setsuya Nakada

Abstract

This chapter sketches an overall volcanic profile of the Japanese Islands and provides a context for the volcanic geoheritage of Japan. Mechanisms and particularities of subduction zone volcanism are discussed, and the recent volcanic history of the Japanese Islands is compared with the volcanic histories of Chile and Indonesia. It becomes clear that Japan has not experienced very large eruption activities compared to Chile and Indonesia despite having geological similarities. This quiescence possibly implies a statistical likelihood of major eruptions in the near future due to accumulation of magma. It is also shown that the actual scale of volcanic eruptions is a poor determinant of the human casualty; instead, the locations of eruptions (distance from residential areas and access) and level of preparedness or vulnerability of the affected population are important factors. The chapter argues that although it is possible to provide probable eruption scenarios, accurate detailed forecasting remains difficult, as each volcano is a different system and the eruption style is not always identical even at a single volcano. It is also argued that fundamental research on individual volcanoes is indispensable to understand this dynamic earth heritage, and reflecting on experience of geoparks in Japan, the chapter states that such heritage branding could become effective tools for promoting awareness and resilience of local societies.

Keywords

Volcanic heritage • Volcano distribution • Disaster vs. scale • Volcanic hazards and geoparks

3.1 Introduction

Volcanoes are among the most fundamental parts of the natural heritage of Planet Earth because of their beauty and elegant appearances, combination with unique ecosystems, and their dynamic behaviors. 21 of the 32 National Parks in Japan are related to active volcanoes. Volcanoes were worshipped as deities in ancient societies for their symbolic landscapes and awe-inspiring shows of force that sometimes generated disasters. The word volcano is derived from the Roman god Vulcan who is said to use a bellow in a fire workshop inside one of the Mediterranean volcanoes, sending smoke and fire from the mountain's chimney. Another famous volcano goddess is Pele of the Hawaiian mythology, who is said to live in the Halemaumau crater of Kilauea volcano, presiding over fire, thunder, dance, and violence. She is beautiful but jealous, reflecting both beautiful-spectacular and violet-destroying behaviors that can be commonly attributed to volcanoes.

In the Japanese mythology, Izanami is the goddess of volcano. She is depicted as a goddess of fertility and love, living under the ground. She bore the islands of Japan within the ocean. According to Hotate (2012), similar mythology about the origin of islands exists in the Pacific Ocean region

S. Nakada (✉)
Earthquake Research Institute, University of Tokyo, Tokyo, Japan
e-mail: nakada@eri.u-tokyo.ac.jp

A. Chakraborty et al. (eds.), *Natural Heritage of Japan*, Geoheritage, Geoparks and Geotourism,
DOI 10.1007/978-3-319-61896-8_3

corresponding to the Circum-Pacific Volcanic Belt. The first recorded volcanic eruption of Mount Aso, found in an old Chinese document from mid-seventh Century, describes a fire pillar that stood from the crater and reached the heaven. Local people enshrined the mountain as they thought the act was divine. This shows that people's fears against unavoidable natural power had governed the society even during the eighth–ninth Century of Japan (Hotate 2012).

In modern times, we have come to possess the knowledge on the formation of volcanoes, how volcanic eruption occurs, and how we can prepare for hazards form such eruptions. In this chapter, volcanological knowledge and the difficulty of eruption forecasting is introduced to argue for informed coexistence with such dynamic earth heritage and to highlight the necessity of preparing for volcanic hazards.

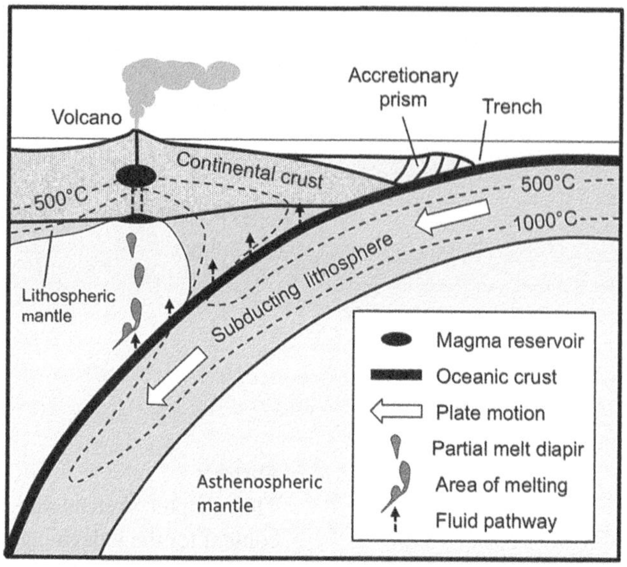

Fig. 3.1 Schematic cross section of subduction zone where the oceanic plate descends under the continental plate. The oceanic plate consists of the oceanic crust and the lithospheric mantle, while the continental plate consists of the continental crust and the lithospheric mantle. Thermal contour lines are from Wade and Wang (2009). The zone for melting in the asthenospheric mantle is defined by the combination of the pressure and temperature with the amount of water component added from beneath; inferred just above ~100 km depth of the subducting oceanic crust

3.2 Why Volcanoes Exist in Japan

The reason behind the existence of a large number of volcanoes in Japan is that the Japan Arc has geologically developed under collision of continental and oceanic plates. In the Japan Arc, oceanic plates interact with parts of continental plates and "sink" under the latter in a process called subduction. The continental plates in question are parts of the Eurasian continental plate and the North American continental plate which were separated from continental Asia by the back-arc spreading of Sea of Japan about 21–15 Ma (Otofuji et al. 1985). Due to the long history of subduction encompassing more than 400 million years (Kojima et al. 2016), the Japanese Islands have grown toward the Ocean side by accretion of sediments scraped off from the surface of the oceanic plates (Fig. 3.1). Strain that accumulates along the boundary between the continental crust and the subducting oceanic crust, or within the continental crust for a long time, can be released to generate sudden slips of the boundaries or faulting of the strata within the crust, respectively. Therefore, earthquake activity occurs so long as plate subduction continues. The surface of the oceanic plate sinking into the mantle is hydrous and becomes the site of melting due to high temperature in the mantle. Alternatively , the sinking oceanic crust squeezes out water components as its dips into the high-pressure mantle. Partial melting of the mantle due to decreasing melting temperature by the addition of water produces magma in this overlying mantle (Fig. 3.1). Magma produced in this process rises and accumulates under the crust due to the density contrast between the mantle and the crust. The magma continues to rise and accumulates in the gravitationally neutral level of the middle crust, forming a magma chamber (e.g., Grove 2000). During volcanic eruptions, this magma rises to the surface. Basically, similar to earthquakes, so long as the oceanic plate subduction continues, magma production continues.

From this discussion, we can understand how restless seismic and volcanic activities at the plate boundaries drive land formation in the Japanese Islands. Major landforms of the Japanese Islands had been created through accretion of sediments and deposition of ejecta of volcanic eruptions, and those landforms were repeatedly deformed by dynamic movement of plates and resultant volcanism.

3.3 Volcano Distribution in the Japanese Islands

The distribution of volcanoes is controlled by the geometry of plate subduction (Fig. 3.2). In NE Japan, volcanoes are distributed nearly parallel to the Japan Trench where the Pacific Plate subsides under the NE Japan. The group of volcanoes belonging to this distribution is called the East Japan Volcanic Belt. On the other hand, in SW Japan, volcanoes are distributed nearly parallel to the Nankai Trough, and the distribution is called the West Japan Volcanic Belt. These places, where the depth of the subducting plate surfaces reaches about 100 km (Pacific Plate for the

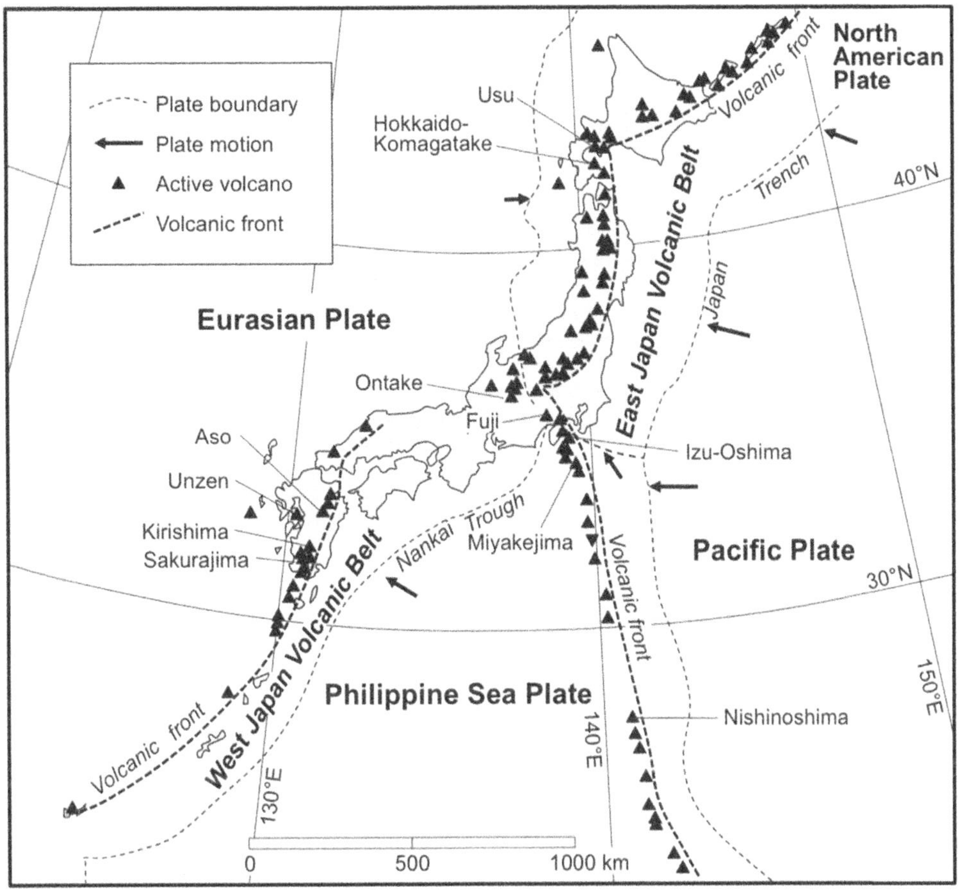

Fig. 3.2 Distribution of active volcanoes in the Japan Archipelago. The lengths of black arrows represent the relative motions of plates. Data from Nakada et al. (2016) and JMA (2013)

East Japan Volcanic Belt and the Philippine Sea Pate for the West Japan Volcanic Belt), correspond to the margin of the volcano distribution to the ocean side, which is also known as the "volcanic front." In each volcanic belt, the density of volcanoes grows toward the volcanic front and the magma production rate is highest just below the position of the volcanic front. Volcanoes with large-scale eruptions are located on the front (Nakada et al. 2016). Large-scale eruptions are commonly associated with the formation of caldera: a large depression resulting from the collapse of the unsupported rock body over the magma chamber. The magma chamber itself becomes nearly vacant for hours to days as a large amount of magma is taken out during the eruption. In the Japanese Islands, caldera-forming eruptions occurred in Central and Southern Kyushu, Hokkaido, and Northern Tohoku (Fig. 3.3). It is considered that the caldera nesting areas correspond to areas where the strain rate of the crust is the least; i.e., magma could steadily accumulate without interruption by frequent eruptions in these locations (Takahashi 1995). In the last 150,000 years, caldera-forming eruptions took place 14 times, and the statistical probability of VEI 7 eruptions is as low as about 0. 1/1000 year (Nakada 2015).

3.4 Characteristics of Volcanism in Japan

It can be said that in terms of volcanic activity, the Japanese Islands are currently in a quiescent phase. To understand this quiescence, it would be better to start with understanding the scale of volcanic eruptions. In volcanology, the scale of eruption is described by either the Volcanic Explosivitiy Index (VEI) or the magnitude of eruption (M). The former is based on the volume of *tephra* (a Greek word for eruption products issued from the crater into the air, including volcanic ash, block ejecta including lava fragments, pumice and scoria) from one explosive eruption, while the latter is based on the mass (weight) of all products from one eruption. Both indices range from 0 to 8. The VEI is an indicator reflecting the size of area impacted (covered) by tephra, and it can also be estimated roughly from the height of eruption column standing above the crater, which is related to the areal dimension of tephra dispersion. In case of eruptions where lava is flowing or is piled up above the crater, such as in Kilauea of Hawaii and Unzen of Japan, respectively, VEI values are small but the M values are intermediate, because only a

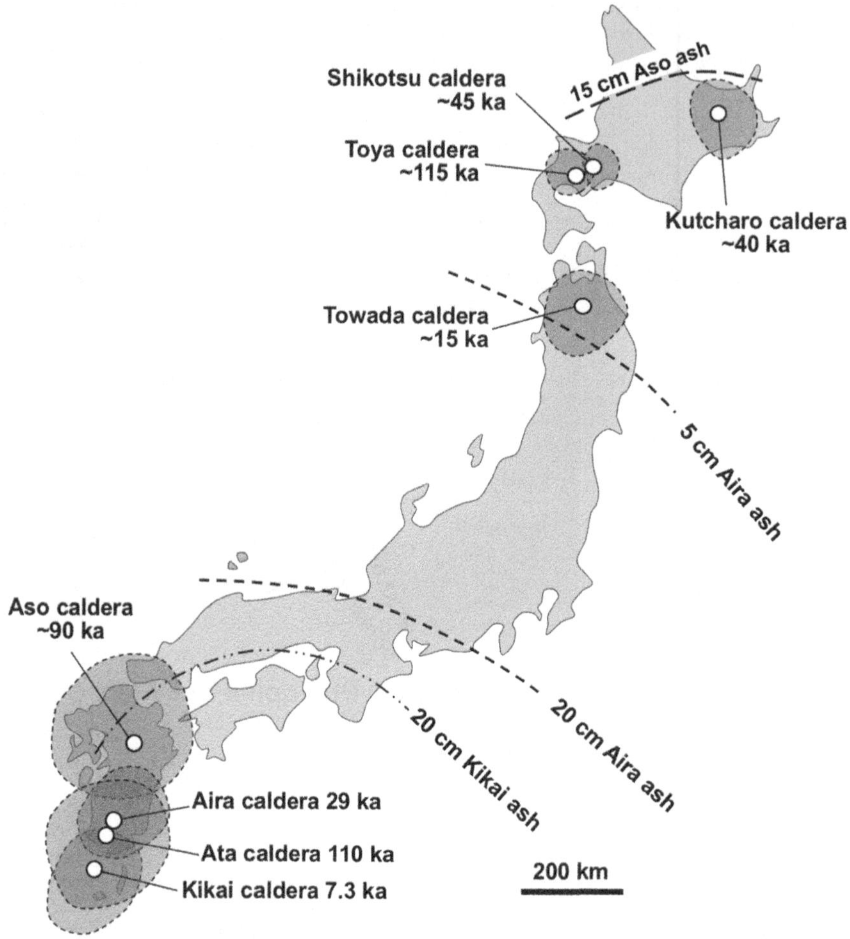

Fig. 3.3 Distribution of representative caldera volcanoes and the contours of main volcanic ash layers from caldera eruptions. "ka" means thousand years ago. Gray circular areas around caldera volcanoes are estimated as signature of pyroclastic flows associated with caldera eruptions. Modified from the National Research Institute for Earth Science and Disaster Resilience, http://dil.bosai.go.jp/workshop/02kouza_jirei/s18kasairyu/f6caldera.htm (accessed on August 30, 2016). Data from Machida and Arai (2003), JMA (2013), and Nakada et al. (2016) were referred

minor amount of tephra is ejected during those eruptions. However, when the scale of eruption is large enough, such as a caldera-forming event, both indicators show similar values.

General description and volume of tephra of eruptions can be listed below:

Non-explosive eruption	VEI 0	$<10^4$ m^3 of tephra
Small eruption	VEI 1	10^4–10^6
Moderate eruption	VEI 2	10^6–10^7
Moderate Large eruption	VEI 3	10^7–10^8
Large eruption	VEI 4	10^8–10^9
Very large eruption	VEI 5 and >5	$>10^9$

Caldera-forming eruptions normally have VEI of 6–8. The largest eruption for 150,000 years in Japan occurred at Mount Aso of 90,000 years ago, the volume of ejecta was >600 km^3, and the explosivity was rated at rank VEI 7 (Machida and Arai 2003). The latest VEI 8 eruption was recorded in Northern Sumatra of Indonesia 75,000 years ago; the Toba caldera was formed by the eruption with about 2800 km^3 (2.8×10^{12} m^3) of magma (Rose and Chesner 1987). Eruptions ranked VEI 7 and 8 are called "super eruptions."

In Fig. 3.4, the VEI values of eruptions in Japan during the past 350 years are compared with eruptions in Indonesia and Chile. Both countries have many active volcanoes like Japan. Geological backgrounds of volcanism are also similar, typified by subduction-related volcanism of the same Circum-Pacific Volcanic Zone. Figure 3.4 clearly shows that eruptions VEI 5 and more never occurred in Japan for these 300 years; in contrast, such eruptions are common in both Indonesia and Chile. In Japan, one of two last VEI 5 eruptions was the Hoei eruption at Mount Fuji in 1707. Furthermore, it becomes obvious that eruptions of VEI 4 never occurred during the last 100 years, another clear contrast to the situation in Indonesia and Chile. Geologically, there is no reason to believe that the recent geological background of recent Japan is any different from that of two countries, so the quiescence in Japan is accidental and it might suggest accumulation of magma under Japanese Islands that has not been spent by large-scale eruptions.

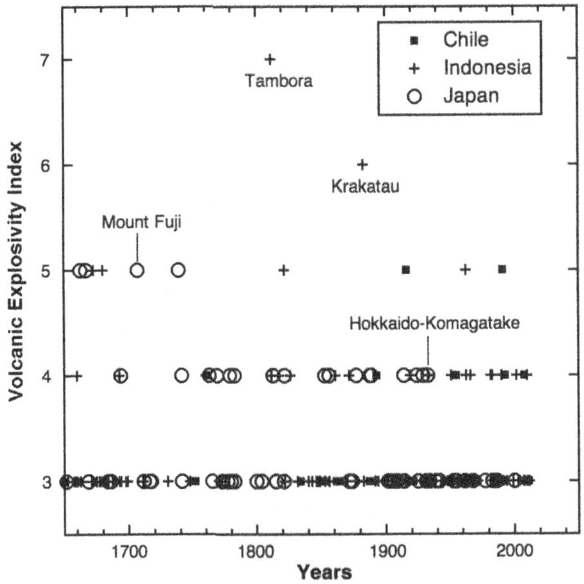

Fig. 3.4 Comparison of Volcanic Exlposivity Indices of volcanic eruptions among Japan, Indonesia, and Chile for these 350 years. Data after Nakada (2015)

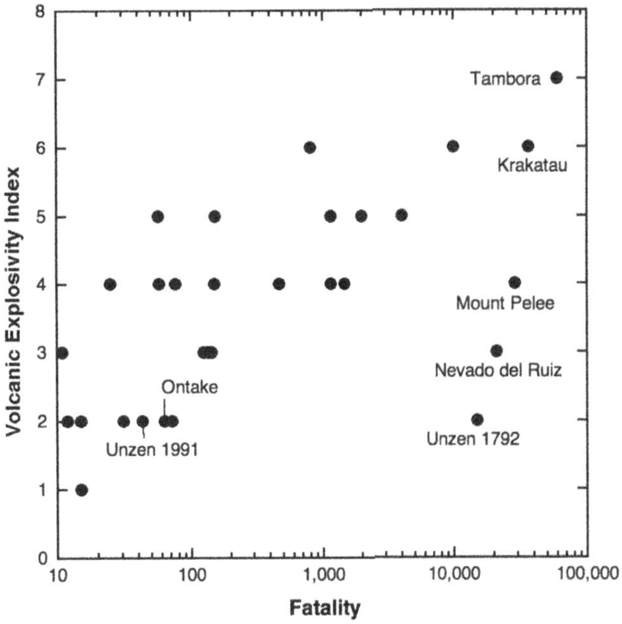

Fig. 3.5 Volcanic Explosivity Indices *vs.* numbers of fatalities for volcanic eruptions in Japan for these 300 years and for major eruptions in the world

3.5 Disaster vs. the Scale of Eruption

A small phreatic eruption at Mount Ontake (3067 m. asl) on September 27, 2014, caused a fatality count of 63 (including missing people). This is the largest number of casualties in a volcanic disaster after World War II. Mount Ontake is one of the famous volcanoes that have nurtured a cult of mountain worship, and the mountain is a popular destination for climbers as it is easy to climb; a cable car takes hikers to 2150 m. The day of the event (Saturday) also fell in the beginning of the fall season when the mountain slopes are covered with colorful leaves. It was a fine day with a blue sky. The eruption occurred on 11:53 a.m., just as hundreds of climbers gathered at the summit to take rest and lunch. The eruption sites were located at a hydrothermal activity site called Jigoku-dani (hell valley) on the southern slope a little below the summit (Maeno et al. 2016b). Most climbers did not initially notice the eruption, which was not accompanied by loud noises of explosions. Climbers took photos of dense cloud arising from below, forgetting to take shelter. Within 30 min, a large amount of rock fragments up to half of a meter across twice fell on the summit area. Climbers except those who had fled to mountain huts were hit by those rock fragments. Fortunately, people in the huts (about 100 of them) could escape the impact of falling stones. The VEI of this eruption was 2.

As shown in Fig. 3.5, the scale of eruption is not related to the scale of damage. The factors that amplify damages from natural hazards include the number and resilience of people who are affected and fragility/robustness of infrastructure (strength of houses and shelters, roads for evacuation, and so on). Even if large eruptions were to occur in a remote area, the human damage would not be large. However, if a large number of people are exposed to natural hazards, such as the climbers on Mount Ontake, the damage becomes large. According to the summary for recorded eruptions by Auker et al. (2013), fatal volcanic events (in terms of lives lost) are more frequent in the eruptions of VEI 2–5 rather than those of VEI >5, though fatal events with the largest total damages are also from large-scale eruptions. This result implies that fatal events occur within a moderate distance (a few to 10 km) from the source, where people can live and approach the source of the hazard. To minimize risks from volcanic hazard, the mind-set of avoiding natural hazards and preparedness for hazards based on proper knowledge on volcanic eruptions is important.

The Aso caldera eruption that occurred around Aso volcano in Kyushu about 90,000 years ago covered the whole area of Japan with fallout; the thickness of ash was >15 cm in Hokkaido (Fig. 3.3). The analysis for recorded eruptions by Auker et al. (2013) showed the fatal event from tephra (ashfall) is usually limited to <<10 km from the source. Ashfall itself is not critical, if meals and water can be prepared in shelters. The most lethal volcanic hazard is the pyroclastic flow. Pyroclastic flows are a mixture of tephra, hot gas, and ash that can travel as fast as >100 km/h. Areas attacked by pyroclastic flows are completely destroyed, burnt, and buried by thick ash and lava fragment deposits. In the case of the caldera eruption at Mount Aso 90,000 years ago, pyroclastic flows reached about 100–150 km in all directions from the source (Fig. 3.3). In the case of the Kikai caldera eruption, south of Kyushu Island, about 7000 years ago, a part of the Jomon-era habitation in the southern part of Kyushu disappeared due to pyroclastic flows crossing the sea. Tsunami generated by collapse of volcanic islands or pyroclastic flows entering the sea also causes heavy damage. The largest volcanic disaster in recorded history of Japan is a Tsunami disaster at Unzen Volcano in 1792 (Fig. 3.5) which recorded a fatality count of 15,000. This event was marked by the collapse of an old volcano triggered by a large earthquake and a large amount of collapsed debris rushed into the Ariake Bay resulting in the generation of Tsunami waves attacking both shores.

3.6 Examples of Recent Eruptions

3.6.1 Shinmoedake Volcano

The Shinmoedake volcano in the Kirishima Volcanic Group erupted magma in January 2011 after a dormancy of 300 years (Nakada et al. 2013; 2016). Columns of volcanic ash repeatedly formed up to 7 km above the crater (Fig. 3.6). Large volumes of pumice and ash from the eruption plume were carried by the northwesterly wind and covered the area near the crater and the residential area southeast of the volcano. After strong explosion events of the first 2 days, lava appeared at the summit crater, grew as a lava dome, and filled the floor. Due to sealing of the magma vent by the lava dome, another type of explosion occurred in February that shattered the lava dome. Eruptive activity continued until early September 2011 with intermittent explosions. The eruption of large quantities of pumice associated with tall plumes of volcanic ash was the first in japan in several decades, the last such event being the 1977 eruption at Usu Volcano. Again, this is another evidence for a quiet phase of volcanic activity in recent years in Japan. The total volume of tephra from this eruption was about ~ 1×10^7 m^3 and that of new lava within the crater was about 1.5×10^7 m^3). The total mass of this eruption reached about 5.5×10^7t; the scales were 3 as VEI and 3.7 as M.

Fig. 3.6 Photos of eruptions in Shinmoedake Volcano in 2011. (**a**) A Vulcanian explosion in the afternoon of January 27, 2011. Photo taken about 3 km south of the crater. (**b**) Smoke from the center of the flattened lava dome within the crater of Shinmoedake. In the background, Takachiho and Ohachi volcanoes are visible. Photo taken on February 2, 2011

Shinmoedake is in the territory of the Kirishima Japanese Geopark. A volcano hazard map was prepared under the Council of Trans-Kirishima, under the guidance of the science advisor of the geopark, and information was distributed through the geopark to all homes around the volcano before the 2011 eruption. During the eruption, up-to-date and accurate information on the eruption including a quick and efficient evacuation plan was provided to the community through the geopark. Disaster prevention education for school children was also carried out effectively and geopark people joined the volunteers to remove volcanic ash from rooftops.

3.6.2 Nishinoshima Volcano

A volcanic eruption occurred near the Nishinoshima Island in the Pacific Ocean in 2013 (Maeno et al. 2016a). The uninhabited island is located about 1000 km south of Tokyo. In this island, an older eruption had occurred about 40 years ago, but large sections of the newly formed land were subsequently lost to wave erosion. In November 2013, eruptive activity started about 300 m southeast of the shoreline of the Nishinoshima Island with an explosive submarine eruption from the shallow sea (around the depth of a few tens of meters). Black-colored cock's tail jets were repeatedly issued from the sea surface, due to the magma's interaction with seawater (phreatomagmatic eruption). About a week after the explosion, the head of magma (lava) reached above seawater, causing the termination of phreatomagmatic activity and the eruption style changed into non-explosive lava flow with repeated bursts of lava from the central crater (Strombolian-type eruption). During the expansion of lava, the old island (Nishinoshima) was nearly swallowed by the new lava land (Fig. 3.7). Lava effusion with the Strombolian eruption continued in the summer of 2015. Late in November 2015, the eruption of Nishinoshima came to an end. The total volume of ejecta was estimated about 1.3×10^{8} m^3 including the materials erupted under the sea . This corresponds to M 4.3 (but VEI ~ 0).

3.7 Difficulty of Forecasting Volcanic Eruptions

Forecasting volcanic eruptions continues to be at the forefront of volcanological research for the last 40 years. In Japan, many scientists are involved through different national projects. As a result, we are now able to detect signals before most of the eruptions, when monitoring is carried out by multiple geophysical and geochemical methods. Especially for those volcanoes that are repeating eruptions in recent years, we can simulate the most likely scenarios of the future eruptions. However, abnormality signals are different in one volcano from the other; indeed, indicator signals are not always similar even in a single volcano. Most eruptions from recent years differed from the previous eruptions at the same volcano. Thus, it is clear that very complex mechanisms govern volcanic eruptions, and even if the approaching eruption can be detected, it is difficult to forecast the eruption scenarios (development) after the actual onset (including the manner of eruption, scale, and duration).

One suggestive example can be seen in the Mount Ontake eruption that was discussed earlier (Maeno et al. 2016b). The last eruption before 2014 occurred in 2007; the inflation of the summit area was observed with the escalation of seismicity during that event. During September 10–11 in 2014, a seismic swarm occurred under Mount Ontake, and the Japan Meteorological Agency (JMA) issued the Volcanic Information that indicated a possibility of a small phreatic eruption. However, seismicity declined in the following days, leaving only a few low-frequency seismic events, and no detectable inflation of the summit was observed except for the last few minutes prior to the eruption. The initial information on the seismic swarm itself was not accepted by the local administrative units as the danger sign, and consequently, no information was given to the climbers on that day. Thus, this event marked an instance where forecasting failed to generate adequate response, and this failure effected the reconsideration of volcanic disaster-related information and its handling by administrators and scientists.

Volcanoes commonly repeat similar phenomena from the past. Furthermore, eruptions do not occur randomly; instead, they follow a crude statistical rule whereby larger eruptions are less frequent than smaller eruptions, and the magma release rate for a volcano or a volcanic region is roughly constant over long temporal scale (Nakada 2015). Based on these facts, we may be able to ascertain the probability of future eruptions for a volcanic region. As every volcano is a distinct system, continuation of scientific research on every active volcano to understand these relationships and patterns is vital.

3.8 Preparing for Volcanic Hazards and Geoparks

From the foregoing discussion, it is clear that nearly all regions of the Japanese Islands are likely to be affected by a major volcanic eruption at some point, and many locations continually experience small- to moderate-scale volcanism. Therefore, volcanoes and their dynamics are fundamental aspects of the geological heritage of this archipelago. In this sense, Japanese volcanologists and science educators have a unique responsibility not only to continue research at the forefront of volcanic eruption forecasting, hazard mitigation, and monitoring of volcanoes but also to reach out to people and disseminate information in an effective manner. As the majority of Japanese geoparks feature volcanic themes, geoparks have emerged as a key tool for disseminating volcano-related information. However at the same time, volcanoes are complex systems and no two eruptions are completely the same. This necessitates more fundamental level research on each volcano and understanding the diversity of volcanic mechanisms. It also requires a change in the mind-set and risk awareness level of the local administrators and residents. Geoparks can play effective roles in achieving such aims.

Due to the natural variation of volcanoes and the socio-economic differences of concerned areas, preparing for volcanic eruptions varies from administrative to local resident levels. Monitoring with the help of the newest technology in

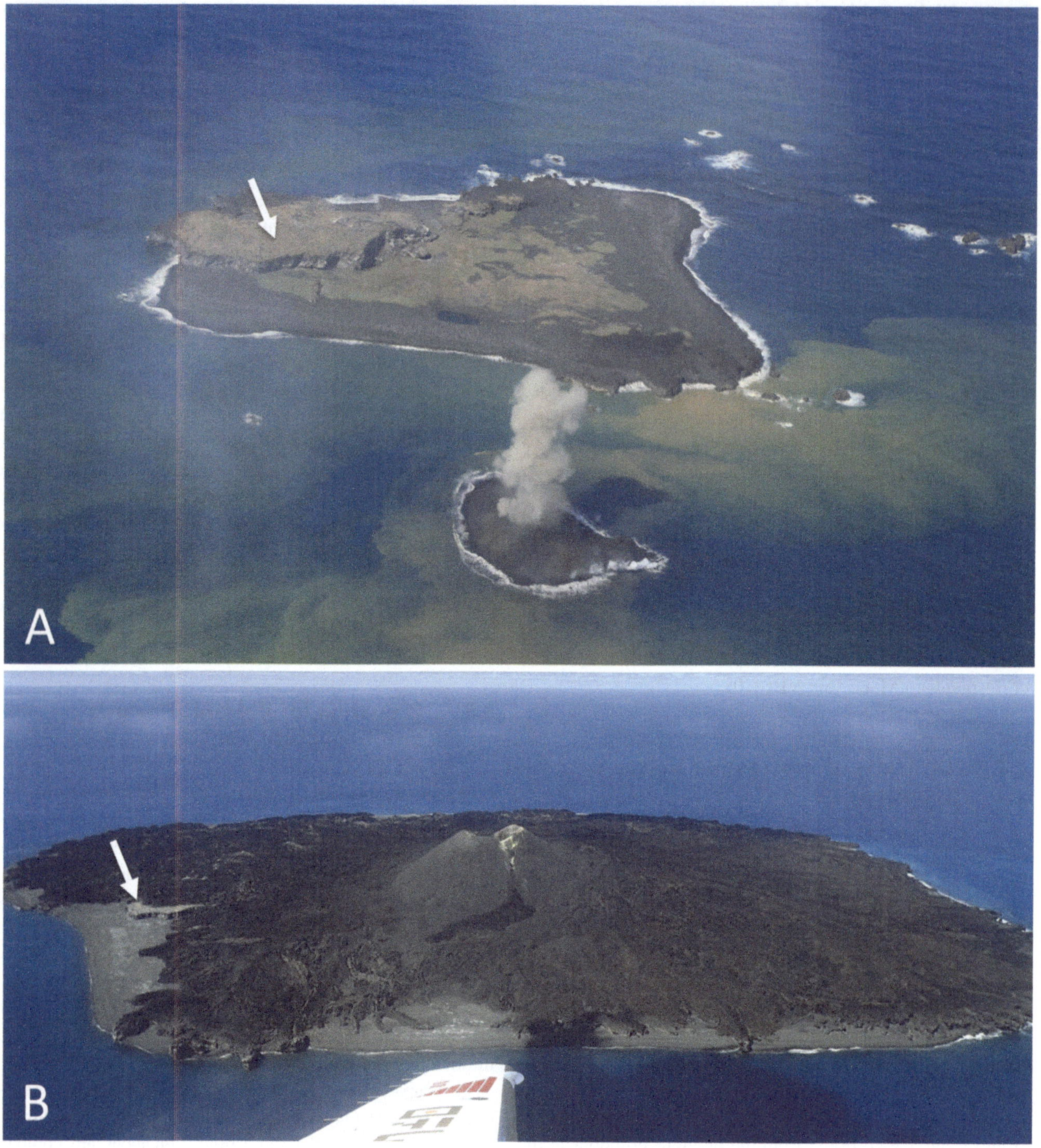

Fig. 3.7 Photos of eruptions in Nishinoshima volcano. (**a**) Eruption that commenced offshore of Nishinoshima started creating new volcano island. Photo November 24, 2013. (**b**) After the end of eruption, a new volcanic island grew and covered the older island with lava flows. Photo July 2, 2016. White arrows indicate the same cliff of the older Nishinoshima Island in the beginning of and after the 2013 eruption. The eruption center is represented by a scoria cone taller than the surrounding lava flows. Both photos were taken from aircrafts of the Asahi and Yomiuri Shimbuns, respectively

multiple ways and at multiple places is important for active volcanoes. Establishing information systems on and around volcanoes is important as well. This in turn will facilitate providing information to local people and climbers. Geoparks can be a medium where monitoring data released by the JMA are transferred and explained in easy-to-understand terms; this will increase the potential of swift response in case an emergency occurs. Planning resilient infrastructure for volcanic disasters is also important; this includes setting of shelters and signboards to show routes for evacuation near craters and availability of disaster (or hazard) maps. Exercises (drills) for evacuation can involve the local people and can raise awareness of volcanic phenomena and the role of hazard maps. Administrators can also realize weaknesses of planning and receive guidance on volcanic crisis management in the process. As the community becomes an important unit in case of volcanic crisis, role plays of both community leaders and members should be discussed before and during exercises. Elevating the knowledge of people including climbers is another *must* when coexisting with volcanoes. Such knowledge is critical for saving lives on the mountains. In a natural disaster, the extent of the damage depends on the resilience or vulnerability of the people. The points mentioned above can raise the resilience of local people and tourists, effecting an increase of resilience in societies, communities, and individuals in the face of volcanic hazards, and can foster the potential to recover quickly from damage.

There are two excellent UNESCO Global Geoparks: the Unzen Volcanic Area and Toya Caldera-Usu Volcano, which contain highly active volcanoes of Unzen and Usu, respectively. The last eruption at Unzen (1990–1995) was a lava dome-forming eruption that lasted for four and half years. Principal casualties were due to pyroclastic flows generated by partial collapse of the lava dome. On Unzen volcano, a lava dome that formed in the last eruption has become an excellent natural monument of the volcanic landscape and a memorial for volcanic disasters. This is also an interesting case where a new geological heritage was formed by a very recent eruption. The community that experienced this volcanic disaster increased their resilience, and currently, geopark guides in this area showcase the potential of the community to respond to future volcanic disasters. The Usu volcano repeats eruptions every 20–50 years within the Toya Caldera that formed about 110,000 years ago (Fig. 3.3). The last eruption in 2000 was started by a small pumice eruption, and the event was followed by phreatic (water-laden) eruptions. Soon after the eruption, the "Volcano Meister" system was introduced in the geopark; this program has trained guides to convey knowledge to visitors and to the next generations on the history and the hazards of the volcano, based on their experience of disaster and recovery.

3.9 Conclusions

In ancient societies, volcanoes inspired awe as gods or goddesses, and people could do little except praying when faced with volcanic disasters. Today, modern technology enables us to diagnose volcanoes and understand anomalies before eruption events. Although eruption mechanisms have been studied by scientists, understanding such mechanisms is not always enough for forecasting eruption onset and scale. Timely forecasting remains difficult and predicting detailed scenarios (development of eruption) is even more difficult. However, probable scenarios of the future eruptions can be provided based on geological research on the eruption history of individual volcanoes. Such research also has the additional important function of improving hazard maps that can be utilized for city planning and evacuation during crises. As this chapter argued, the scale of volcanic disasters is not directly dependent on the scale of eruptions. The exposure of people and their properties is more important as a factor. Although recently Japan is abnormally quiet in terms of volcanic activity, large eruptions are a constant possibility. It is in this sense that more fundamental research on each active volcano is required, especially for a land that has been borne directly out of subduction and volcanism in an active plate margin . Not only does such research help us understand the beauty of volcanoes, the complexity of their behavior, and the diversity of the heritage they engender, but such efforts can also be effectively translated to a more resilient and better informed society through tools such as geoparks and other heritage labels.

References

Auker MR, Sparks RSJ, Siebert L, Crosweller HS, Ewert J (2013) A Statistical analysis of the global historical volcanic fatalities record. J Appl Volcanol 2:2. http://www.appliedvolc.com/content/2/1/2. Accessed 31 Aug 2016

Grove TL (2000) Origin of Magmas. In: Sigurdsson H et al (eds) Encyclopedia of volcanoes. Academic Press, New York, pp 133–170

Hotate M (2012) Rekishi no Nakano Daichidouran: Nara Heian no Jishin to Tenno (Earth disturbance in the history–earthquake and the Emperor of Nara and Heian). Iwanami Shinsho, Tokyo

Japan Meteorological Agency (JMA) (2013) National catalogue of the active volcanoes in Japan, 4th edn. Japan Meteorological Agency. http://www.data.jma.go.jp/svd/vois/data/tokyo/STOCK/souran_eng/menu.htm. Accessed 31 Aug 2016

Kojima S, Hayasaka Y, Hiroi Y, Matsuoka A, Sano H, Sugamori Y, Suzuki N, Takamura S, Tsujimori T, Uchino T (2016) Pre-Createaceous accretionary complexes. In: Moreno T et al (eds) The Geology of Japan. Geological Society of London, London, pp 61–100

Machida H, Arai F (2003) Shinpen Kazanbai Atorasu: Nihonrettou to Sono Shuhen. Atlas of Tephra in and around Japan, Revised edn. The University of Tokyo Press, Tokyo

Maeno F, Nakada S, Kaneko T (2016a) Morphological evolution of a new volcanic Islet sustained by compound lava flows. Geology 44 (4):259–262. doi:10.1130/G37461.1

Maeno F, Nakada S, Oikawa T, Yoshimoto M, Komori J, Ishizuka Y, Takeshita Y, Shimano T, Kaneko T, Nagai M (2016b) Reconstruction of a phreatic eruption on 27 September 2014 at Ontake Volcano, Central Japan, based on proximal pyroclastic density current and fallout deposits. Earth Planets Space 68:82. doi:10.1186/s40623-016-0449-6

Nakada S (2015) Regularity of volcanic eruptions in terms of Volcanic Explosivity Index (VEI). Bull Volc Soc Jpn 60:143–150 (In Japanese with English abstract)

Nakada S, Nagai M, Kaneno T, Suzuki Y, Maeno F (2013) The outline of the 2011 eruption at Shinmoe-dake (Kirishima), Japan. Earth Planets Space 65:475–488

Nakada S, Yamamoto T, Maeno F (2016) Miocene-Holocene volcanism. In: Moreno T et al (eds) The Geology of Japan. Geological Society of London, London, pp 273–308

Otofuji Y, Matsuda T, Nohda S (1985) Opening model of the Japan Sea, inferred from paleomagnetism of the Japan Arc. Nature 317:603–604

Rose WI, Chesner CA (1987) Dispersal of ash in the Great Toba Eruption, 75 ka. Geology 15:913–917

Takahashi M (1995) Large-volume felsic volcanism and crustal strain rate. Bull Volc Soc Jpn 40: 33–42 (In Japanese with English abstract)

Wada I, Wang K (2009) Common depth of slab-mantle decoupling: reconcileing diversity and uniformity of subduction zones. Geochem Geophys Geosyst 10:Q10009. doi:10.1029/2009GC002570

Part I

World Heritage Sites (Natural)

Shiretoko Peninsula: Dynamic Interaction Between Geology, Geomorphology, and Ecology at The Interface of Terrestrial and Marine Systems

4

Abhik Chakraborty

Abstract

This chapter describes and analyzes the Shiretoko Peninsula World Heritage area as a complex and dynamic system comprising of geological, geomorphological, and ecological characteristics. Declared a national park in 1964, Shiretoko became a World Heritage Site in 2005. The area is noted as an exceptional example of the interaction between marine and terrestrial environments, as the most southerly location of drift sea ice in the Northern Hemisphere, and as an environmental system that links tiny diatoms to whales and brown bears. Several endangered and iconic species such as the Blakiston's Fish Owl and the Steller's Sea Eagle can be seen here, and the surrounding waters support many fish species and large marine mammals such as the Steller's Sea Lion, the Orca, and the Sperm Whale. However, the area has witnessed landscape fragmentation and change in ecosystem dynamics in the past due to anthropogenic impact and is also currently threatened by Global Environmental Change. This narrative evaluates Shiretoko as a combined and complex geo-ecological system, emphasizing the complexity, uncertainty, and plurality of the interaction between its different components as fundamental properties that have implications for its management as well.

Keywords

Shiretoko Peninsula • Complex system • Combined geo-ecological system • Endangered species • Landscape level process • Fragmentation

4.1 Introduction: The Shiretoko Peninsula World Natural Heritage, An Outline

The Shiretoko Peninsula is a jutting peninsular landmass stretching nearly 70 km into the Pacific Ocean (Sea of Okhotsk) in Northeast Hokkaido. Located on the Chishima (Kurile) arc, the peninsula bears a history of submarine and terrestrial volcanism that goes back nearly 9 Ma. While the base of the peninsula is covered by submarine volcanic depositions, the topography is dominated by uplift and Quaternary volcanic processes driven by the magma produced as the Pacific Plate subducts under the continental margin (North American Plate). It has been noted that the Chishima (Kurile) arc shows episodic explosive volcanism (Prueher and Rea 2001), and while the magnitude of volcanic events is currently significantly reduced, the area still features active volcanoes such as Mt. Rausu (1661 m asl, the highest point of the peninsula) and Mt Io (1563 m asl). Rapid uplift and volcanic depositions on the narrow land (land gets narrower closer to the tip) have resulted in a steep topography that did not allow human habitation or sedentary activities (Natural Parks Foundation 2015), which in turn has contributed to the wild and diverse nature of Shiretoko persisting to this day. In addition, the area is affected by heavy snowfall from November to April, resulting in relative restriction of human activity (MOE 2012) (Fig. 4.1).

A. Chakraborty (✉)
Center for Tourism Research (CTR), Wakayama University, Wakayama, Japan
e-mail: abhik78@hotmail.co.jp

A. Chakraborty et al. (eds.), *Natural Heritage of Japan*, Geoheritage, Geoparks and Geotourism,
DOI 10.1007/978-3-319-61896-8_4

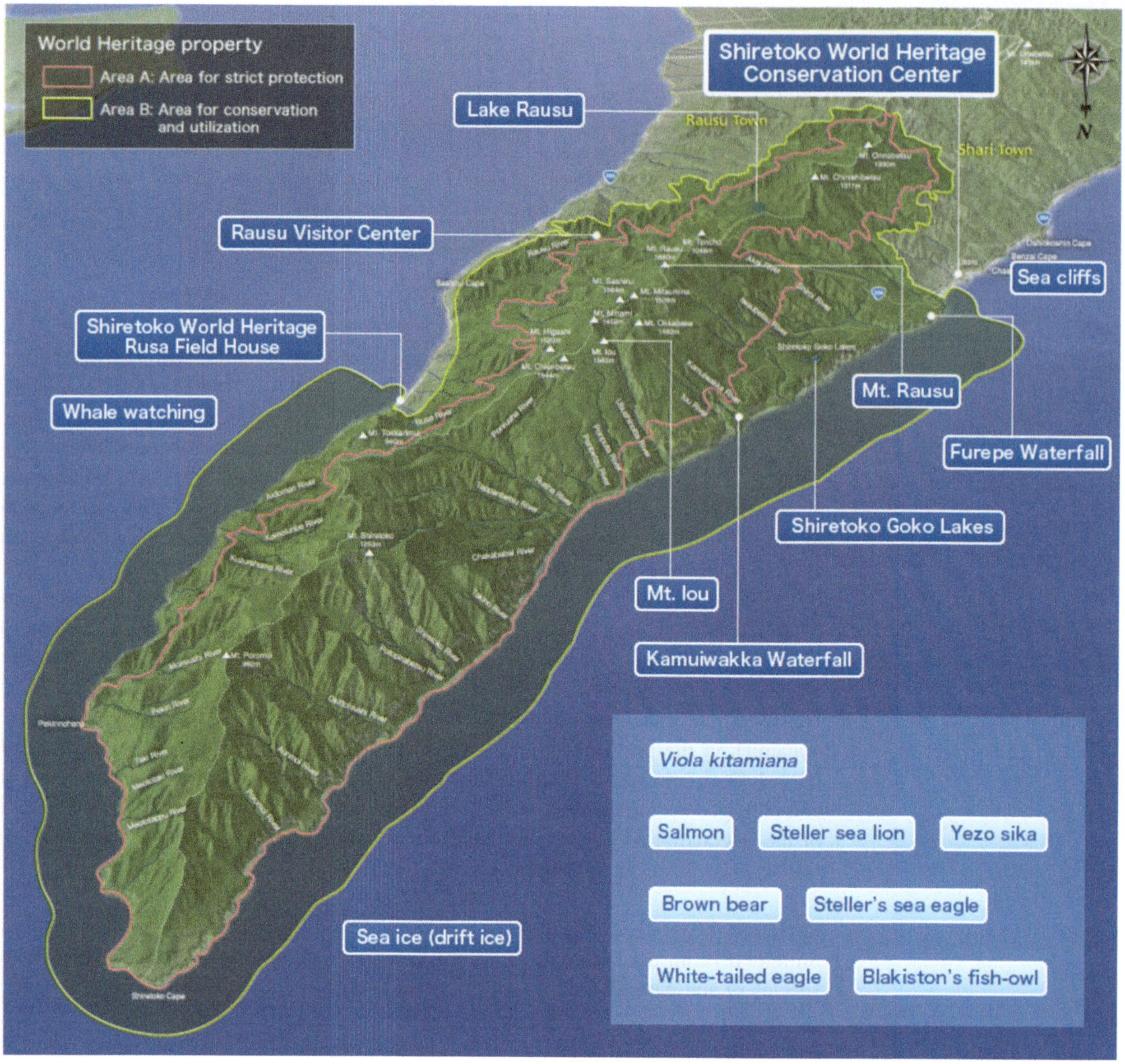

Fig. 4.1 Map of Shiretoko Peninsula World Heritage. Source: Ministry of the Environment website (World Natural Heritage in Japan) (https://www.env.go.jp/nature/isan/worldheritage/en/shiretoko/area/index.html) (accessed on 2017/07/13)

The Shiretoko National Park came into existence in 1964; it was a late addition to the National Park list in Japan. The Ministry of the Environment mentions that the main reason for inscribing the territory onto the National Parks list was the protection of the wild landscape: currently, the 38,636 ha National Park has nearly 60% of the land registered as "Special Protection Zone" and no amount of land falls under "Common Zone" (allowing human activities with minimum restriction) (MOE 2012; Natural Parks Foundation 2015). In this regard, Shiretoko National Park is unique among Japanese national parks, most of which were created with the promotion of some form of tourism activity in mind. According to the MOE (2012), ecosystems support large mammals such as the Orca (*Orcinus orca*) and the Brown Bear (*Ursus arctos*) and endangered species such as the Blakiston's Fish Owl (*Ketupa blakistoni*); mutual feedback between drift (sea) ice, marine environment, and natural forests; landscapes of remarkable beauty shaped by volcanism and erosion; and a local society that lives using the bounties of nature and protecting the same are the four main characteristics of Shiretoko National Park (Fig. 4.2).

Shiretoko was inscribed onto the UNESCO World Heritage (Natural) List in 2005 (property size of 71,100 ha including a marine zone makes the WHS larger than the

Fig. 4.2 (**a**) Weathered cliffs reveal mechanisms of volcanism in Shiretoko (Photo by author). (**b**) Iwaobetsu Rivermouth from air, showing the formation of sea ice around a peninsula dominated by volcanic mountains. (Photo copyright Shiretoko Nature Foundation, reproduced with permission)

national park territory). The terrestrial part of the property is distributed between the Shari and Rausu Towns. The main reasons for inscribing the property and therefore the Outstanding Universal Value (OUV) are an outstanding example of interaction between terrestrial and marine ecosystems (Shiretoko is described as one of the best examples of integrated ecosystems on the WHS list); the formation of drift (sea) ice at the lowest (most southerly) latitude in the Northern Hemisphere; nutrient-rich seawater of the Sea of Okhotsk underpinning a diverse food web based on a rich phytoplankton production that in turn supports numerous marine mammals such as cetaceans and sea lions; and endemic and endangered species such as the *Viola kitamiana*, the *K. blakistoni*, and salmonid species that spawn in the rivers (UNESCO nd). The WHS inscription raised the value of the area to a global level, and thus, evaluating the natural heritage, the challenges it faces, and possible solutions become a matter of global significance.

Apart from being a major tourist attraction of the area, drift ice is a special geomorphological feature that sustains the Shiretoko ecosystems. The formation of the sea ice is due to a complex mechanism of marine currents, topography, and atmospheric circulation. In their seminal paper on the formation of drift ice, Ohshima et al. (2001) describe how continental drainage (input of nutrient-rich freshwater from the Amur River) creates prominent stratification at 50–100 m of mean depth that suppresses winter convection at greater depths, leading to the formation of drift ice which is then advected southward under the combined influence of sea (the East Sakhalin current) and atmospheric (northerly winds) circulation (Ibid, Hiwatari et al. 2008). Sea ice provides ideal conditions for growth of phytoplankton that forms classical algal bloom in spring when the drift ice melts. This attracts a range of zooplankton, which in turn attracts larger species of predators, sustaining a vast cycle that includes the microscopic algae and species at the top of food chains such as the Orca and the Seller Sea Lion in the waters, the Brown Bear on the land, and the Steller's Sea Eagle roaming the sky. This functioning web of geo-ecological interactions sustains the natural heritage and its more touristic components of Shiretoko. Some key species and their interaction with the system components are described below.

How the continental shelf topography and oceanic circulation combine to create the unique geo-ecological conditions of the area, and that of the broader Oyashio Large Marine Ecosystem (spanning nearly 530,000 km^2), is described in detail by Sakurai (2007). The Shiretoko system therefore is a part of the larger Sea of Okhotsk and Oyashio (literally meaning "parent current" for its productivity) system (Ibid). Sakurai (2007) observes that the system could be in a state of *flux* or *regime shifts* over long timescale by demonstrating temperature fluctuations in relatively recent years and consequent change in the biomass. Fluctuations occurring over much longer time scale are described by Kawahata et al. (2003) where they describe oscillations in the southern Sea of Okhotsk in the Holocene. As the system provides considerable benefits to local residents, the economic repercussion of such change is also keenly felt.

This chapter focuses particularly on a "systemic appraisal" of the Shiretoko Peninsula, albeit from a mainly terrestrial point of view. The terrestrial landmass of the World Heritage Area provides an excellent body of evidence of how marine and terrestrial ecosystems interact and how, at the species and habitat levels, anthropogenic change impacts the system functions or exacerbates the influences of Global Environmental Change (GEC). The chapter analyzes Shiretoko as a complex adaptive and combined geo-ecological system and explores the stress on various system components. The observations are deemed significant for the conservation and management of this dynamic and complex natural heritage.

4.2 Data Collection and Analysis Method

This chapter is based on a review of key literature and empirical data gathered during two field trips in the territory. The field trips were conducted in order to confirm leads gained from the literature review and further explore angles that were deemed significant. During the fieldwork, local stakeholders such as National Park officials, visitor center officials, tourism operators and guides, and specialists (scientists) working on different facets of the Shiretoko ecosystem were interviewed. Identification of current threats faced by the property was done based on two important compilations brought out by the Yomiuri Shimbun (2006) and Shiretoko Museum (2010), which provide an exhaustive review of environmental issues.

As this chapter provides an explanation of the *characteristics* of the Shiretoko system, it ventures to do so from an angle of the *qualitative aspects* of this dynamic system. While complex adaptive systems and complex ecological systems have been documented and analyzed from quantitative points of view that typically focus on the interaction(s) between a limited number of variables, that approach needs to be complemented with analysis of the inherent complexity of these systems that are fundamentally irreducible, with appraisals that are synthetic, and with appreciation of the fundamental uncertainty and plurality of knowledge itself (Stirling 2010). Most available analyses on complex systems such as Shiretoko, to my knowledge, are based on reductive knowledge that seeks to clarify questions on uncertainty by showing clear causal relations. However, as today's increasingly complex and *wicked* problems show, such reductive analysis is not enough for addressing complex environmental problems. Thus, the analysis provided here consciously attempts to look at the holistic picture by focusing on conduits of connectivity in the system and aims to draw *connections* across categories because, in reality, the value of this natural heritage is based on a yet poorly understood complex web of feedbacks between its different components. The lessons from this chapter are revisited in the Synthesis chapter to draw broad conclusions about multifaceted conservation and governance of natural heritage.

4.3 Landscape Level and Ecosystem Level Characteristics of Shiretoko

The Shiretoko Peninsula, as described above, is characterized by volcanism and uplift that resulted in a steep terrain dominated by volcanic mountains. There are 871 species of plants in the area out of which 97 are endangered, including species such as the *V. kitamiana* and *Carex stylosa* C.A. Mey that are endemic to this region

(Ishikawa 2010). A good description of the zonation of vegetation can be seen in the vicinity of the 1661 m Rausudake, where 0–500 m elevation areas feature species adapted to coastal cliffs and mixed forests of conifers and broadleaf species; the 500–1000 m zone is marked by species such as the *Betula ermanii* and *Alnus Maximowiczii* and the region above 1000 m typified by mixed patches of *B. ermanii* and *Pinus pumila,* and *P. pumila* colonies (MOE 2012). On windswept mountainsides that are unsuitable even to the hardy *P. pumila*, uniquely adapted shrub colonies persist, and snow-patch communities are observed in locations that are covered by snow for most of the year. *V. kitamiana*—often cited as a plant symbolizing the unique ecological conditions of Shiretoko—is suited particularly to rocky high-altitude locations. Thus, at every level of the landscape, a strong feedback between the prevailing geomorphic conditions and plant adaptation is observed (Fig. 4.3).

The peninsula has many drainage channels due to heavy snowfall, but rivers are short and rapid and they flow with a steep gradient and high bedload. Nearly 90 moderate and small streams cascade down from highlands to the sea in Shiretoko (Fig. 4.4), sometimes plunging into the sea as waterfalls. Under natural conditions, such rivers perform the two vital tasks of actively eroding and shaping the landscape (carrying the mountains to the sea) and supporting a high natural variation of species some of which straddle marine and terrestrial ecosystems (such as fish species that swim upriver to spawn and swim downriver to the sea after hatching). The rivers of Shiretoko are known to be the southernmost limit for the sea run of Dolly Varden as well as a variety of Pacific salmonids (Chum, pink, masu Salmon, and White-spotted Charr are some notable species) (UNESCO nd). These species support mammals such as the Brown Bear (*U. arctos*) and endangered avian species such as the Blakiston's Fish Owl (*K. blakistoni*). While the former's hibernation cycle is closely associated with the migration of the salmonids, the latter is nearly completely reliant on rivers and valley forest compositions for habitat. In addition, biomass derived from the fish that end their life cycle in the upstream and from the excreta of their predators produce vital nutrients for the forests which in turn support both predator and prey species. Currently, rivers in the peninsula are significantly affected by anthropogenic impact: dams, check weirs, artificial embankments, and channel engineering hamper the flow of material along the channels and the functioning of the river corridors. The Yomiuri Shimbun (2006) compilation mentions how the Iwaobetsu River drastically changed in the 1960s after check weirs prevented any significant upstream migration of the masu salmon species *Oncorhynchus masou masou*.

Shiretoko is known for its Brown Bears, a species that is found all over Hokkaido (Fig. 4.5). The omnivorous Brown Bear feeds on an extremely broad range of diet, and although 90% of its diet is based on plants, it can occasionally be seen scavenging carcasses of large animals such as the *Cervus nippon yesoensis* (Ezoshika Deer) and cetacean species. The estimated range of activity of a Shiretoko individual is confined within 15 km^2 (Natural Parks Foundation 2015). While this is a fraction of the more extensively roaming habits of its North American counterparts, this characteristic is quite possibly the result of adaptation to a habitat shaped by geomorphological conditions and aided by the ample food source provided by those conditions.

The Blakiston's Fish Owl (*K. blakistoni*), considered to be the largest owl species in the world, is another iconic species. Large individuals can attain a body length of nearly 70 cm and a wingspan of nearly 2 m. The Shiretoko area is considered as the habitat for nearly half of all Blakiston's Fish Owl population of Hokkaido (Takenaka 2010). Currently, the species is adversely affected by a lack of large trees (of over 1 m diameter at breast height) and the decline of fish in the rivers (due to damming and other engineering inputs in the river valleys) (Ibid), but a moderate increase of numbers has been reported, indicating success of the fish owl conservation program (Fig. 4.5).

Finally, Shiretoko is an important place for migratory birds and avian species that are adapted to the terrestrial–marine interface. Every year in spring, the Short-tailed shearwater (*Puffinus tenuirostris*) arrives in large flocks from the Southern Hemisphere, crossing vast distances to feed on the zooplankton that thrive on the phytoplankton bloom. The sight of these small birds diving into the waves is considered symbolic of spring in Shiretoko. In the winter, one of the largest species of eagles, the Steller's Sea Eagle (*Haliaeetus pelagicus*) arrives in search for food in the relatively warmer Shiretoko from its reproductive habitat in the Kamchatka region (Fig. 4.6). Currently, this majestic species is considered vulnerable in the IUCN list of threatened species. The White-tailed Eagle (*Haliaeetus albicilla*) is another familiar bird of prey. The Spectacled Guillemot (*Cepphus carbo*) finds its reproductive habitat in the rocky ledges that define the coastline of the peninsula (MOE 2012).

The description above is a bare outline of some of the most iconic species found and observed in the Shiretoko World Heritage area, mainly pertaining to the terrestrial environment. The marine environment around the peninsula supports many more iconic and endangered species such as the Sperm Whale (*Physeter macrocephalus*) Orca (*O. orca*), Steller's Sea Lion (*Eumetopias jubatus*), Dall's Porpoise (*Phocoenoides dalli*), Spotted Seal (*Phoca largha*), and Ribbon Seal (*Phoca fasciata*)—many of these species are inadequately studied to date, implying much uncertainty remains on their habitat conditions, their interactions with a changing environment, and their population dynamics.

a

b

Fig. 4.3 (**a**) Looking toward Mt. Rausu from a wintry day reveals snowcover on the highlands (Photo by author). (**b**) Hemerocallis dumortieri (C. Morren) var. esculenta (Koidz.) Kitam. ex M. Matsuoka and M. Hotta near Mt Rausu. Mountains of the peninsula are known for rich alpine flowers. (Photo copyright Shiretoko Nature Foundation, reproduced with permission)

Fig. 4.4 Short and rapid rivers coming out of mountains are a major landscape feature of this area. However, many river courses are artificially altered, hampering movement of species and biomass. (Photo by author)

Fig. 4.5 (**a**) A Brown Bear (*U. arctos)* grabbing a pink salmon (Photo by Takeshi Takenaka). (**b**) Blakiston's fish owl (*K. blakistoni*) is the largest owl species of the world (Photo by Takeshi Takenaka)

4.4 Tourism Activities in Shiretoko World Heritage Area

Shiretoko Peninsula and the waters off it count as a major ecotourism and wildlife-watching destination in Japan. Nearly, 2.5 million tourists visit the peninsula every year (Natural Parks Foundation 2015). In the two administrative units of Shari Town and Rausu Town, tourism foci are contrasting: while in Shari most tourist activities are centered around terrestrial attractions, the fishing town of Rausu offers tourism packages—especially those in the winter—with a marine environment-related theme.

In Shari, the Five Lakes of Shiretoko (Shiretoko Go-ko) form a notable tourist attraction. The area offers a scenic view of the volcanic mountains, a pleasant walking experience in the forests and around the five lakes that formed due to large-scale landslide, transport, and deposition of volcanic products. There are several viewpoints to observe and understand the major land features of the peninsula, and there is a wooden walkway that can be used from spring to autumn. This walkway, completed in 2010, is an example of good

Fig. 4.6 A Steller's Sea Eagle (*H. pelagicus*) (Photo by author)

practice and innovative engineering. The walkway is barrier free and is fenced with an electric hedge that allows visitors to walk freely in a territory that is also a natural habitat of brown bears. Electricity for fence protection is generated by solar panels at the parking area (Natural Parks Foundation 2015). While constructing the structure, care was taken not to cut down any tree in the location, a series of poles are used to hold the entire structure above ground (Ibid). While it cannot be used in the winter, the walkway almost certainly contributes to minimizing the effect of trampling of the vegetation during peak seasons. In winter, snowshoe tours are organized in the area, and these tours have gained popularity over the years. Visitors can enjoy the view of the frozen Furepe Waterfall, and wildlife such as Ezoshika Deer and Red Fox (*Vulpes vulpes* schrencki) walking on the snow covered landscape, and on clear days, these experiences can be complemented by a panorama of snowy mountains. Crossing the peninsula via the trans-Shiretoko road is another popular sightseeing activity from late spring to autumn. In late spring and early summer, "walls" of snow are seen along the roadside. The road allows visitors to begin their trip at Shari and end at Rausu or do it the other way around. The road is closed in the winter due to heavy snow. The Shari area also has a number of scenic waterfalls (Fig. 4.7).

Sea kayaking is popular along the Shari coast. As most of the core area is out of bounds for tourists, kayaking allows visitors to get a close view of the coastal topography and forests. Kayaking packages in nearshore areas are available in winter as well. Alternatively visitors can enjoy coastal landscape and wildlife from sightseeing boats.

In Rausu, the focus of tourism is mainly on marine wildlife. All year round, cruise packages that offer glimpses of cetaceans, other large marine mammals such as seals and sea lions, and migratory birds (mainly in winter) are available. Some cruise packages can be combined with staying in the facilities operated by the same company. In the winter, cruise tours are particularly popular among birdwatchers aiming to photograph rare species such as the Steller's Sea Eagle. A lodging facility named the "Fish Owl Observatory" is also popular since it offers chance to see and photograph the elusive Blakiston's Fish Owl. Enjoying hot springs (including some secluded ones) and local seafood delicacies and hiking to the Rausu Lake are some other notable tourism activities in this area.

In addition, hiking the volcanoes of Shiretoko is popular in summer and autumn. The 1661 m tall Rausudake can be hiked from both Shari and Rausu Towns. Hikers walk through diverse natural forests and wildlife in the lower part of the hiking course. The upper section of the trail offers chance to see a number of alpine flower species, including the rare *V. kitaminana* [which can be seen in the vicinity of Mt. Iozan (1562 m)]. The 1547 m Sharidake is another popular hiking destination (the trailhead is located in the nearby Kiyosato Town). While none of the hikes require advanced mountaineering skills and nor are they particularly physically demanding, visitors must be aware that there is no toilet facility beyond the trailhead. Visitors should carry

Fig. 4.7 (**a**) The frozen Furepe waterfall is a major tourist attraction in the winter (Photo by author). (**b**) Mt. Rausu in summer (Photo by Takeshi Takenaka)

portable toilets and should refrain from relieving themselves in the wild or discarding tissue papers in order to protect the sensitive mountain ecosystem.

As a World Heritage area, Shiretoko possesses excellent visitor centers at multiple locations with knowledgeable staff. The World Heritage Center of Shari (Utoro Visitor Center of the Shiretoko National Park) is conveniently located at the road station *Shirietoku* at the gateway of the Utoro area (where most of the lodging facilities are concentrated). The center features an excellent and interactive exhibition, committed staff, a variety of pamphlets and information brochures, and a small library, though most documents are available in Japanese only. The Shiretoko Nature Center, managed by the Shiretoko Nature Foundation, serves as the second visitor center of Shari with its knowledgeable staff and exhibits. It is also a hub of wildlife management experts. This facility is open all year round (excepting the new year break). The World Heritage and National Park Visitor Center of Rausu is another excellent facility to learn about the nature of Shiretoko: the visitor center has exhibits of an Orca skeleton and stuffed specimens of a number of other important animal and bird species in addition to thematic panels that guide the visitor to different characteristics (geology, geomorphology, ecology, and landscape) of the area (Fig. 4.8).

As the description above shows, the Shiretoko area has strong credentials as a World Heritage as it offers good quality nature experience packages and adequate facilities to explain about the natural heritage of the area. Adequate management expertise is also available. Due to these factors, the area typically elicits a comparatively high degree of satisfaction among its visitors and a significant ratio of visitors express their desire to visit the area again; a fuller account of those aspects is available in Romao et al. (2014). However, there are some challenges associated with tourism as well, and a brief description of those challenges and their implications for natural heritage management is provided below and some points are followed up further in the analysis.

A major problem is the conflict of wildlife and tourists. There is an apparent increase of brown bears venturing close to tourists, possibly lured by scraps of food thrown by tourists (Shiretoko Nature Foundation nd), and there is a consequent pressure to chase off or even shoot and kill bears in some cases. The Shiretoko Nature Foundation is currently trying to educate tourists to behave responsibly and stop feeding wildlife (Ibid), noting that the responsibility to avoid accidents fall on the visitors. It is difficult to gauge to what extent the rising popularity of the destination creates more pressure on managers to take proactive measures (in this case disturbing the bear in its natural habitat), but the issue deserves continued attention.

The feeding problem is arguably even more pronounced on the avian species. Particularly in the Rausu area, wildlife cruise boats throw fish to attract birds of prey such as eagles. Shiraki (2010) has observed that the problem has become serious to the extent that most of the birds visible to the birdwatchers are dependent on this artificial feeding, which is subject to tourist demand of the resource (birds in this case). This problem is further discussed below (Fig. 4.9).

Additionally, there are problems such as the lack of multilingual communication ability of guides (resulting in their relative inability to communicate with foreign tourists), discarding of garbage in sensitive mountain environments, lack of visitor knowledge about vulnerable vegetation (particularly in the high-altitude areas), and a perceived lack of interest among tour operators to interact with scientists and conservation experts (Takenaka 2010; personal correspondence). Tour operators could also incorporate visits to the excellent visitor center facilities into their tour packages and educate their clients about the natural environment. But perhaps the most serious issue is the lack of systematic tourism impact monitoring by tourism service providers.

Fig. 4.8 The Shiretoko World Heritage Visitor Center in Rausu has excellent exhibits such as a skeleton of an Orca and stuffed specimens (photo by author)

Fig. 4.9 Birds flocking to a remaining ice floe in search of food thrown from sightseeing boats. Such artificial feeding has repercussions for the natural ecosystem (photo by author)

4.5 Heritage Management Nature Conservation Efforts at Shiretoko: An Overview

The Shiretoko Peninsula World Heritage site is managed by a diversity of stakeholders including the national ministries such as the Ministry of the Environment (MOE) and the Ministry of Agriculture, Forestry and Fisheries (MAFF), local administrative units (Shari and Rausu Towns), the Shiretoko Nature Foundation, and fishery cooperatives. The decision-making structure is supported by a scientific council comprising of natural resource management specialists. Shiretoko is evaluated as a good example of co-management with local stakeholders; Makino et al. (2009) and Sakurai (2010) outline a marine resource management plan which they define as the "Shiretoko Approach." The plan entails autonomy of the local fisherfolk, selective fishing, voluntary restraint measures including limitation of fishing vehicle number and type, and fishing zones and seasons, and it is aided by scientific advice when applicable. The authors also point out that the local fisherfolk possess rich traditional knowledge of the marine environment, which can be utilized for scientific decision-making. However, it is somewhat unclear how "adaptive" the structure is, as Makino et al. (2009) observe that it does not specify numerical targets at species level, nor is it clear to what extent the measures taken by the fisheries are in coordination with other potential stakeholders such as river management specialists (due to the fact that rivers in the peninsula are vital conduits linking terrestrial and marine ecosystems) and wildlife conservation experts.

As for the terrestrial environment, a major turning point of landscape level conservation came in 1977 when the "100 Square Meter Movement" for replanting forests began. The initial goal was to protect areas that were affected by deforestation from construction and development pressure, by reforesting them (Natural Parks Foundation 2015; Shiretoko 100 Square-Meter Movement Trust nd). The main current goal of the project is to protect forests and help increase old-growth forest cover; and the movement breaks this task down in terms of 5-year (short-term), 20-year (mid-term), and 100-year (long-term) timescale goals (Shiretoko 100 - Square-Meter Movement Trust nd). The projects are aided by donations from individuals, groups, and corporations, and the reforestation activities are supported by forestry and forest ecology experts. Expert knowledge is an important facet of this movement: for example, while *Larix kaempferi* is used as a pioneer species for reforestation, plans are set to reduce the total area covered by this species under longer term scenarios.

A related conservation program is the habitat regeneration and reproduction aid project for the Blakiston's Fish Owl in Hokkaido. This program started in 1984 (Natural

Parks Foundation 2015), and two key focus areas of the project are provision of artificial nests for the endangered species in the light of the fact that large trees suited for nesting have declined and measures to prevent electrocution deaths by erecting protective fencing as such accidents are a main cause of untimely death of grown individuals (Takenaka 2010). These projects are carried out by dedicated scientists and voluntary participants and are supported by national level nature conservation programs and private nature conservation initiatives (Ibid; Takenaka 2016 personal interview with author). In the initial stage, the main aim was to protect the fish owl at an individual level. In the past two decades this effort has shown signs of steady success, and the focus of conservation activities is now on both protection of individuals and their breeding environment and on eventually regenerating a diverse natural forest cover that allows the species to exist without threat (Takenaka 2010). This point is followed up in the next section.

The Shiretoko Nature Foundation, established in 1988, is the most prominent stakeholder of wildlife management and forest ecosystem conservation in the area. The organization advocates the need to conserve the integrity of wilderness in the area, noting that the Shiretoko Peninsula is a rare location in the Japanese Islands where large predators such as bears can roam freely across the landscape (their natural habitat), and this wild nature is of fundamental value for this region (Shiretoko Nature Foundation nd). As of 2015, the Shiretoko Nature Foundation undertook 64 projects of conservation and wildlife management with a total budget of more than 338 Million JPY (Ibid).

Thus, the Shiretoko area possesses several significant nature conservation examples, the significance of which extend beyond the local dynamics. Whether it is about protecting keystone species, regenerating forests, or integrating resource users such as fisherfolk into conservation and environmental management, the lessons from these ongoing programs will undoubtedly be valuable for natural heritage management at the global level.

Heritage in a Nutshell: Shiretoko Peninsula
Location: Shari and Rausu Towns in NE Hokkaido
Type of Heritage: Composite Geo-ecological (terrestrial and marine ecosystems of exceptional diversity and value underpinned by geological and geomorphic processes), coastal landforms, volcanic landforms, and unique geomorphological conditions such as the drift (sea) ice formation.

The property is evaluated highly for the connectivity of terrestrial and marine ecosystems that link tiny diatoms to whales (in the sea) and to the Brown Bear on the land. Shiretoko is habitat for a large number of endangered species of plants and animals such as the *Viola kitamiana* plant and the Blakiston's Fish Owl (the largest owl species in the world). The Steller's Sea Eagle (the largest eagle species in the world), the White-tailed Eagle, and the Spectacled Guillemot are notable bird species. The seas around the peninsula support a large variety of fish and large marine mammals including the Sperm Whale, the Orca, the Steller's Sea Lion, and several types of seals. The rivers are known for supporting several salmonid species.

Drift ice formation is both visually spectacular and ecologically significant, as it drives the process of algal blooms, which in turn sustains the higher order components of ecosystems.

Shiretoko is a place where the connectivity between geological, geomorphological, and ecological systems is conspicuous. The Shiretoko National Park highlights the intrinsic value of "wild nature" and commits to protect the natural environment. Land formation is dominated by uplift and volcanism, and the many rivers act as potent erosion agents. However, over the years drift ice has thinned, leading to the possibility that the system might undergo significant change in the near future.

What to see: The area offers excellent ecotourism and wildlife experience packages all year round. Visiting the "Five Lakes of Shiretoko" is highly popular, as are hiking to the Rausu Lake and hiking up the mountains. Off the coast, activities range from diving, sea kayaking, walking on drift ice (in winter), and wildlife viewing cruise packages. More details are available in Sect. 4 of this chapter. The World Heritage Visitor Centers of Shari and Rausu are also excellent facilities where visitors can learn about the natural heritage of the area.

Useful websites:
MOE Shiretoko Peninsula https://www.env.go.jp/en/nature/nps/park/shiretoko/index.html
Shiretoko Nature Foundation http://www.shiretoko.or.jp/en/

4.6 Challenges for Managing the Natural Heritage: Landscape Level Fragmentation, Anthropogenic Pressure, and Global Environmental Change

Although Shiretoko is often perceived as a "pristine" natural environment, the area witnessed several types of anthropogenic impact. This section provides an outline of some key issues and challenges.

In their authoritative study on forest fragmentation in Hokkaido, Shoyama and Braimoh (2011) note that since 1869 (when land conversion for development projects began in Hokkaido), forests in Hokkaido have declined drastically both in quantity and quality; old-growth forests spanning 4360 km^2 were converted to farmland (22), and citing municipal government data, the authors note that for particular areas such as the Horobetsu-Iwaobetsu plateau, forest loss became drastic in the twentieth century—showing a strong positive correlation with rapid cultivation and land conversion (*kaitaku* in Japanese) that took place between 1945 and 1966, but the trend was somewhat reversed due to reforestation from 1978 (Ibid: 23). Their findings indicate that out of the mixed conifer-broadleaf forests that characterize low elevation sections of the peninsula, broadleaved species showed a consistent net decrease from 1947 to 2004 (time of data collection). A further important insight from the study reveals that natural forest succession occurred in the areas which were abandoned after a phase of deforestation, and the sequence of dwarf bamboo brush pioneering the forestscape, followed by broadleaf species, to the eventual regeneration of broadleaf-conifer mixed canopy, was observed. However, despite natural regeneration and reforestation programs, forests did not recover their pre-cultivation integrity; the mean patch size in 2004 was only 24% of the pre-cultivation mean patch size (28). Thus, there is strong evidence that agriculture and pastureland conversion in the area was a major cause of primary forest fragmentation in this area in the twentieth century and that despite natural succession and planned reforestation, the forest landscape is still significantly fragmented and its ecosystem connectivity compromised to a large extent.

Fragmentation at the landscape level is also indicated from the extensive damming and other forms of engineering that waterways of the peninsula have been subjected to. There are an estimated 247 small and moderate sized dams and flood or silt control structures on the rivers in this area (Takahashi 1999; Morita and Yokota 2002). The Yomiuri Shimbun (2006) compilation also notes that due to extensive check weir construction on the waterways, the land and sea connectivity in the ecosystems was damaged and the effect is particularly serious on the fish species that spend most of their life cycle in the rivers such as the Dolly Varden (*Salvelinus malma*) (Fig. 4.10). The species is adversely affected and is currently threatened. It is worth noting that healthy river fish stocks are vital for keystone species such as the Blakiston's Fish Owl, which are forced to search for food across larger distance when rivers go devoid of fish and can become susceptible to accidental deaths. Thus, from the viewpoint of long-term conservation of an iconic species like the fish owl, both regeneration of large patch size of naturally diverse forests and restoration of nutrient and species conduits that sustain those forests (such as natural river basin mechanisms) are required.

Among the changes occurring spontaneously (without direct anthropogenic impact) to the ecosystems is the overgrazing by the Ezoshika Deer. The Ezoshika spread to Shiretoko in the 1970s from the nearby Akan area (Uno 2010) and subsequently their numbers multiplied rapidly. Although there were natural "breaks" in the form of starvation deaths in the winter (such as a starvation-related decrease of the total population by over 60% in 1999), numbers have recovered quickly to the extent that overgrazing by these animals has become a threat. Particularly worrying is the effect on overgrazing on the Shiretoko Cape area that is home to many endangered and endemic plant colonies; it was reported that colonies of *Empetrum nigrum* var. japonicum, an alpine plant that adapted to the post-glacial environment and has an ecological history of nearly 10,000 years, were decimated by overgrazing at the cape area (Yomiuri Shimbun 2006). Other endangered or sensitive species such as the *Aconitum misaoanum f. album* and *V. kitamiana* could be affected as well (Uno 2010). Currently, the MOE and the Shiretoko Nature Foundation are conducting planned reduction of Ezoshika numbers by selective culling to control overgrazing damage.

Along with such fragmentation of landscapes, pressure brought by urbanized lifestyle and tourist demands on the

Fig. 4.10 A Dolly Varden (*S. malma*); this species is a preferred food of *K. blakistoni*, but human interference in river regimes has resulted in its swift decline in local streams (Photo by Takeshi Takenaka)

natural ecosystem is also significant. During field research, I witnessed the artificial feeding of sea eagles in Rausu, an issue that has potential ramifications at the species and ecosystem levels. Shiraki (2010) analyzed how the sudden and brief increase of Alaska Walleye Pollock (*Theragra chalcogramma*) in the Rausu area attracted large flocks of Steller's Sea Eagle and White-tailed Eagle between the mid-1980s and early 1990s, with an estimated 2500 individuals belonging to these species feeding from fishing remnants. Estimates indicate that nearly 90% of the Steller's Sea Eagle population migrated to Shiretoko in winter at that point (Yomiuri Shimbun 2006). Walleye Pollock fishing proved unsustainable due to the rapid decline of the species in the 1990s, and eagle numbers started to decline as well. However, this decline was somewhat reversed from the late 1990s, when fishing vehicles started to throw fish in order to attract eagles (Ibid) which became a tourist attraction (supplanting Walleye Pollock fishing as income source in some cases). This practice continues to this day. Shiraki's data collected in 2007 reveal that 80–90% of the eagles were concentrated on the artificial breakwaters near the port, clearly showing their dependence on artificial feeding. According to Shiraki, prominent problems that arise in the process are (1) unnatural frequenting of human inhabited areas by eagles leading to increased chance of accidental deaths, (2) vulnerability to pathogens due to unnatural density of birds in small areas, (3) possible change of feeding patterns due to change in tourist demand scenario, and (4) the possibility that species such as the *Larus schistisagus* and *Corvus macrorhynchos* that actually take a large share of the thrown food to multiply and affect the ecosystem (2010: 57–59). It is also documented that a number of Steller's Sea Eagles were victims to lead poisoning as they were attracted to deer carcasses left out by hunters in the late 1980s (Yomiuri Shimbun 2006). Although it can be argued that bird watching currently provides an environment-friendly method of local income generation, visitor pressure amounting to "loving nature to death" (Hall 2015) is a common problem and needs to be addressed by educating visitors about the value of wildlife in its natural habitat. The problem is exacerbated by the aforementioned fragmentation of rivers; with the rivers not able to provide adequate feeding opportunities (fish), sea eagles that migrate to Shiretoko are even more attracted to artificial feeding. This case shows the problem of how positive valuation of wildlife as a tourism resource (or commodity for indirect consumption through activities such as bird watching) can possibly degrade the resilience of natural ecosystems and affect species behavior and adaptation, thereby initiating further change in the role of species in its food web.

Human–wildlife conflicts are probably most conspicuous in case of large marine mammals. From the documentation done by Yomiuri Shimbun (2006), two cases stand out: (1) the culling of the Orca (*O. orca*) and (2) the culling of the Steller's Sea Lion (*E. jubatus*) by fisherfolk. While the Orca was culled in large numbers in the postwar period leading to a speculation that the local population around Shiretoko had become extinct at one point (the species has currently recovered and is a major tourist attraction), the Steller's Sea Lion is culled for causing harm to the local fish catch. A particular problem surrounding marine species is that their ecosystems are poorly understood in comparison to terrestrial ecosystems, implying that it is particularly difficult to understand the nature of anthropogenic pressure on them.

Of late, the most conspicuous change, at least at the level of local perception, has taken place in the dynamics of drift ice formation. Nearly, everyone who was interviewed during the field research phase replied that drift ice characteristics have changed over the past few years, including change in the timing of onset, (possible reduction of) volume, drifting pattern, and abrupt breakup. Such changes appear to persist despite year-to-year variations of the drift ice conditions (i.e., the footprint of Global Environmental Change (GEC) was identifiable above the "noise" of natural variation). Such local opinions tally well with the more extensive field data collected by the Yomiuri Shimbun (2006), where it is indicated that till the 1990s, the drift ice was so thick that boats and trawlers were confined within the ports in the height of winter. A key indicator is the start of the sightseeing trips on the *Aurora*, a 490-ton icebreaker. The very fact that this ship started operating in the 1990s was a testimony to the thinning of drift ice, as small icebreakers would not have been able to navigate thick drift ice that prevailed before the 1990s. Experienced sailors spoke of how the thinning trend continued into this century, and in the words of one diver, the ice has become increasingly "softer" in recent years. While the effect of this change on marine and terrestrial ecosystems cannot be adequately clarified due to deficiency of data, the effects are likely to be far-reaching considering the multiple and complex feedback loops between geological formations, geomorphological processes, and ecological characteristics that together form the dynamic natural heritage of Shiretoko (Figs. 4.11 and 4.12).

4.7 A Systemic Appraisal as a Pathway for Better Management?

This study highlighted the challenge of comprehending a dynamic natural heritage and the pathways of recent changes from the case of Shiretoko Peninsula. Apart from the effects of GEC, a number of drivers of change originate locally, or at least, could be addressed locally in order to achieve more effective and holistic governance of this heritage. A systems appraisal is thought to be useful toward this end.

Fig. 4.11 (**a**) Drift ice around the Shari coast (Photo by author). (**b**) Drift ice in the sea off the Shari coast (Photo by author)

The natural heritage of Shiretoko is an evolving system that is supported by geological processes such as volcanism and uplift, geomorphological processes such as erosion by multiple agents (water and ice in this particular case are most important), and transport of material; together these processes sustain the ecosystems of the area. It was noted how the Shiretoko system is influenced by larger geomorphological processes in action such as the Amur River's drainage, the dynamics of atmospheric and oceanic circulation, and changing temperature. It is important to appreciate the facts that the system involves many variables, that their interaction pathways are poorly defined or even chaotic, and that a large degree of uncertainty will always prevail over quantified interrelationship or knowledge describing this system. Such uncertainty is not something to be frustrated about; it is evidence that the system is complex enough to generate new discoveries. In addition, the existence of "uncertainties" must be perceived for their worth at the planning and management levels, which implies that any new program or activity should be thoroughly checked

Fig. 4.12 Drift ice breaking apart. Over the years drift ice has thinned, raising questions about continuity of vital ecological processes (Photo by author)

with scientific knowledge and monitored when it is implemented. Important insight about this system can be gained from *describing* the system from top-down (landscape level to species level) and bottom-up (from species level to landscape level) and by analyzing its key components and interconnectivity in discrete units of space and time. This particularly applies to activities such as tourism, which brings additional and new inputs (pressure and change) to the system. This approach requires exploring the system with *how* questions: some examples of a top-down description involving *how* questions could be: *how* does the flow of water from mountain to sea connect the species along the way? Or, *how* does the drift ice formation affect the feeding behavior of species and *how* are they connected in that process? It is not enough to describe these relationships based on a single year's data; they must be described *over time* to show additional relationships, or *emergence* in the system. To complement this approach, a bottom-up description using *which* and *how* questions could likely be: *which* species are affected by flowing water or the lack of it and *which* species are affected by shrinking sea ice and *how*? Such species (for operational ease, keystone species such as the Blakiston's Fish Owl and the Steller's Sea Eagle could be chosen for the cases provided in the examples, respectively) will need to be monitored over long temporal scales (considering, for example, a fish owl needs a large broadleaf tree to raise its young, and in deforested patches regeneration of such tree cover to appropriate density for the fish owl population will be determined by the revegetation and recovery processes of mixed canopy and is very likely to be influenced by the patch size). A key theoretical lens that can be used for this purpose is the approach of analyzing species-landscape connectivity through the concept of *trophic cascades* (for detailed explanation, see Terborgh and Estes 2010).

Over the past few decades, theoretical explorations have begun to be matched by planning based on ecological connectivity in Europe and North America; a key approach is through the concept of *ecological networks* (Jongman and Pungetti 2004). Lessons from those projects are very much pertinent for places like Shiretoko. A key practical approach has been to "break down" the concept of connectivity in a local scale by considering a given landscape (could be either natural or cultural) and *de-anthropization* of the landscape (i.e., giving more consideration to the ecological characteristics of that site) and considering how they are connected in space and time in order to improve its interconnectedness and environmental continuity (Pungetti and Romano 2004). For monitoring and ecological network planning, typically four approaches are used. These are the census-based, population viability analysis (PVA), landscape index-based, and the spatial standard-based approaches (Verboom and Pouwels 2004). The census-based approach analyzes species distribution in a given spatial area; the PVA approach focuses on the life history of a species (usually an indicator or a keystone species) in order to bring out parameters of a viable population over time; the landscape index-based approach typically uses GIS data to calculate the extent of fragmentation, connectivity of patches, and habitat area size; and the spatial standard-based approach uses Ecologically Scaled Landscape Index (ELSI) for understanding what spatial characteristics imply for a certain species (Ibid: 57–59). While all of these approaches are pertinent for analyzing natural heritage, the adoption of an ELSI-based spatial standard is particularly instructive as it provides important information on both spatial extent and species life history (i.e., landscape–species interaction). However, one important caveat in this regard is that it should be kept in mind that species life history also "adapts" to spatial change over time, and there is a lag effect between species population and landscape change (fragmentation) (Tillman et al. 1994). The lesson from this observation is that the indices should be evaluated through timescale data, or compared with other areas that have similar geo-ecological characteristics but different degrees of change, if possible.

4.8 Conclusion

This chapter analyzed the Shiretoko Peninsula World Heritage Area as a complex, dynamic, and evolving system involving geological, geomorphological, and ecological components. Shiretoko is one of the last remaining "wilderness" areas in Japan; its unique characteristics include interaction between terrestrial and marine environments, species diversity, presence of numerous endangered and endemic species, and occurrence of drift ice at the most southerly latitude in the Northern Hemisphere. The Shiretoko area is remarkable for the tight feedback loops between its geology, geomorphology, and ecological components; the area shows unique landscape ecological attributes as a result. In addition, it is a system that extends far beyond its immediate geographical boundaries and is coupled with geo-ecological systems in the Sea of Okhotsk and northern Pacific. As a World Heritage, Shiretoko has a unique value also because the area featured nature conservation activities from an early date and is managed through consensus building among local stakeholders and advice by a scientific advisory council. However, although Shiretoko has a rich natural environment, there are several "stressors" on this system such as landscape fragmentation (including retardation of functional properties of rivers that are among the most important geomorphological agents), change of forest ecosystems, anthropogenic pressure on keystone species such as the Blakiston's Fish Owl, Steller's Sea Eagle and the brown bear, and GEC (in the form of an apparent warming trend leading to early fragmentation and loss of drift ice). The chapter outlined these "threats" and argued that a systemic appraisal of the property that integrates species level dynamics with the integrity and continuity of geomorphic processes as the key for long-term governance of such complex adaptive and combined geo-ecological systems.

Acknowledgement I thank Dr. Takeshi Takenaka of the Fish Owl Institute for his support and guidance during the field trips to Shiretoko.

References

Hiwatari T, Shirasawa K, Fukamachi Y, Nagata R, Koizumi T, Koshikawa H, Kohata K (2008) Vertical material flux under seasonal sea ice in the Okhotsk Sea North of Hokkaido Japan. Polar Sci 2(1):41–54

Hall CM (2015) Loving nature to death: tourism consumption, biodiversity loss and the Anthropocene. In: Gren M, Huijbens EH (eds) Tourism and the Anthropocene. Routledge, New York, pp 52–74

Ishikawa Y (2010) Shokubutsu no Genjo to Kadai (Current conditions and challenges for vegetation). In: Shiretoko Museum (ed) Shiretoko no Shizen Hogo (Nature conservation in Shiretoko). Hokkaido Shimbunsha, Sapporo, pp 144–175 (In Japanese)

Jongman R, Pungetti G (eds) (2004) Ecological networks and greenways: concept, design, implementation. Cambridge University Press, Cambridge

Kawahata H, Ohshima H, Shimada C, Oba T (2003) Terrestrial-oceanic environmental change in the Southern Okhotsk Sea during the Holocene. Quat Int 108:67–76

Makino M, Matsuda H, Sakurai Y (2009) Expanding fisheries co-management to ecosystem-based management: a case in the Shiretoko World Natural Heritage Area, Japan. Mar Policy 33:207–214

Ministry of the Environment (MOE) (2012) Shiretoko National Park. MOE, Kushiro (in Japanese)

Morita K, Yokota A (2002) Population viability of stream-resident salmonids after habitat fragmentation: a case study with white-spotted Charr (Salvelinus leucomaenis) by an individual based model. Ecol Model 155(1):85–94

Natural Parks Foundation (2015) Shiretoko: Shiretoko National Park. Natural Parks Foundation Publication, Tokyo (In Japanese)

Ohshima KI, Mizuta G, Itoh M, Fukamachi Y, Watanabe T, Nabae Y, Suehiro K, Watatsuchi M (2001) Winter oceanographic conditions in the Southwestern part of the Okhotsk Sea and their relation to Sea Ice. J Oceanogr 57:451–460

Prueher LM, Rea DK (2001) Tephrochronology of the Kamchatka-Kurile and Aleutian Arcs: evidence of volcanic episodicity. J Volcanol Geotherm Res 116:67–84

Pungetti G, Romano B (2004) Planning the future landscape between nature and culture. In: Jongman R, Pungetti G (eds) Ecological networks and greenways: concept, design, implementation. Cambridge University Press, Cambridge, pp 107–127

Romao J, Neuts B, Nijkamp P, Shikida A (2014) Determinants of trip choice, satisfaction and loyalty in an ecotourism destination: a modelling study on the Shiretoko Peninsula, Japan. Ecol Econ 107:195–205

Sakurai Y (2007) An overview of the Oyashio Ecosystem. Deep Sea Res II Top Stud Oceanogr 54(23–26):2526–2542

Sakurai Y (2010) Shiretoko Sekai Shizen Isan Chiiki no Kaiiki Hozen to Kaiyo Hogo-ku (Conservation of the marine ecosystem including marine protected area in the Shiretoko World Natural Heritage Site). Kankyo Kenkyu (Environ Res) 158:1–8 (In Japanese with English abstract)

Shiraki S (2010) Etto suru Ojirowashi to Owashi no Genjo to Kadai (Current conditions and challenges for Winter Migratory White

Tailed Eagle and Steller's Sea Eagle). In: Shiretoko Museum (ed) Shiretoko no Shizen Hogo (Nature conservation in Shiretoko). Hokkaido Shimbunsha, Sapporo, pp 52–61 (In Japanese)
Shiretoko 100 Square-Meter Movement Trust (nd) http://100m2.shiretoko.or.jp. Accessed 30 Mar 2017 (In Japanese)
Shiretoko Museum (ed) (2010) Shiretoko no Shizen Hogo (Nature conservation in Shiretoko). Hokkaido Shimbunsha, Sapporo (In Japanese)
Shiretoko Nature Foundation (nd) http://www.shiretoko.or.jp/shiretoko/. Accessed 29 Mar 2017 (In Japanese)
Shoyama K, Braimoh A (2011) Analyzing about sixty years of land-cover change and associated landscape fragmentation in Shiretoko Peninsula, Northern Japan. Landsc Urban Plan 101:22–29
Stirling A (2010) Keep it complex. Nature 1029:23–30
Takahashi G (1999) Rivers from the viewpoint of mountain stream fishes. Kagaku (Sci J) 69:1036–1040 (In Japanese)
Takenaka T (2010) Shimafukurou no Genjo to Kadai (Current conditions and challenges for the Blakiston's Fish Owl). In: Shiretoko Museum (ed) Shiretoko no Shizen Hogo (Nature conservation in Shiretoko). Hokkaido Shimbunsha, Sapporo, pp 32–51 (In Japanese)
Terborgh J, Estes JA (2010) Trophic cascades: predators, prey, and the changing dynamics of nature. Island Press, Washington, DC
Tillman D, May RM, Lehman CL, Nowak MA (1994) Habitat destruction and the extinction debt. Nature 371:65–66
UNESCO (nd) Shiretoko. http://whc.unesco.org/en/list/1193. Accessed 30 Mar 2017
Uno H (2010) Ezoshika no Genjo to Kadai (Current conditions and challenges for Ezoshika Deer). In: Shiretoko Museum (ed) Shiretoko no Shizen Hogo (Nature conservation in Shiretoko). Hokkaido Shimbunsha, Sapporo, pp 80–87 (In Japanese)
Verboom J, Pouwels R (2004) Ecological functioning of ecological networks: a species perspective. In: Jongman R, Pungetti G (eds) Ecological networks and greenways: concept, design, Implementation. Cambridge University Press, Cambridge, pp 56–72
Yomiuri Shimbun (Hokkaido Branch Compilation) (2006) Umi to Mori no Dorama Shiretoko (Shiretoko: a drama of sea and forests). Kyobunsha, Sapporo (In Japanese)

Shirakami Mountains: Old-Growth Forests of Siebold's Beech Supporting Biodiversity in a Dynamic Landscape

5

Abhik Chakraborty

Abstract

This chapter analyzes the Shirakami Mountains World Heritage Site as a complex natural heritage based on distinctive geological, geomorphological, and ecological components and their mutual interactions. The property was inscribed onto the World Heritage List (Natural) in 1993, becoming one of the initial entrants from Japan. *Fagus crenata* dominates the crown cover of an old-growth forest here; the area is also noted for a rich diversity of flora and as the habitat of several large mammals, a large number of bird species, and a variety of fish in the rivers. Shirakami is a particularly excellent example of a dynamic heritage landscape that is shaped by rapid uplift, erosion, and heavy precipitation. Although currently protected under the World Heritage convention and several national and local level statutes, the forest came to the brink of serious exploitation in the 1980s when a plan to log off the old-growth beech surfaced, leading to citizen protests, nature conservation advocacy, and the eventual registration of the area as a World Heritage. Issues of landscape level fragmentation due to anthropogenic change and Global Environmental Change add to current issues, especially at the peripheral areas adjacent to the heritage property. The chapter concludes that attention should be paid to geological, geological, and ecological connectivity in order to protect the integrity of this important natural heritage.

Keywords

Shirakami Mountains • Old-growth forest • *Fagus crenata* • Connectivity • Fragmentation

5.1 Introduction: Outlines of the Shirakami Mountains World Heritage Site

Shirakami-Sanchi, or Shirakami Mountains, was one of the two initial World Heritage (Natural) registrations from Japan in 1993.[1] With an area of nearly 170 km^2, the area falls under IUCN Criteria Ib,[2] corresponding to a large relatively unmodified wilderness landscape. Shirakami Mountains are noted as the last remaining large forest tract of the Siebold's beech (*Fagus crenata*)—a cool temperate species that spread rapidly at the end of the Last Glacial Maximum (LGM)—in Japan. This old-growth forest managed to withstand natural simplification processes in the post-LGM; it supports a large amount of biodiversity and represents one of the last relatively intact old-growth beech forests of East Asia (UNESCO nd); these are some notable ecological characteristics that contribute to the Outstanding Universal Value of the property. Shirakami Mountains are thus a rare old-growth forest in mainland Japan which retains a largely intact ecosystem due to physical

[1] The other property in scribed as natural heritage in the same year was Yakushima Island.

[2] For more information on IUCN's Protected Area (PA) categories, refer to https://www.iucn.org/theme/protected-areas/about/protected-areas-categories

A. Chakraborty (✉)
Center for Tourism Research (CTR), Wakayama University, Wakayama, Japan
e-mail: abhik78@hotmail.co.jp

A. Chakraborty et al. (eds.), *Natural Heritage of Japan*, Geoheritage, Geoparks and Geotourism,
DOI 10.1007/978-3-319-61896-8_5

features such as the forbidding topography (steep hills) and lack of tracks in the core area (Ibid). But, as I shall discuss in this chapter, it is a landscape that once came to the brink of serious degradation due to a proposed logging project, and the area is currently threatened by loss of forest connectivity and fragmentation of the system components due to anthropogenic impact in the peripheral areas.

The UNESCO World Heritage homepage notes that the beech forests span about 1/3rd of the Shirakami Mountain Range in Aomori and Akita Prefectures in northern Honshu (Fig. 5.1), supporting a diverse range of flora dominated by the "climax" vegetation of *Fagus crenata*; several large mammal species such as the Japanese Black Bear (*Ursus thibetanus japonicus*), the serow (*Capricornis crispus*), and the Japanese macaque (*Macaca fuscata*); 94 avian species, several of which are locally threatened such as the Black Woodpecker (*Dryocopus martius*), the Mountain Hawk-Eagle (*Nisaetus nipalensis*), and the Japanese Golden Eagle (*Aquila chrysaetos japonica*); as well as nearly 2000 species of insects (UNESCO nd; MOE nd). While the reasons for the inscription are mainly biocentric (revolving on Criterion ix of the World Heritage Convention), the area also offers an excellent example of a combined geoecological heritage due to its geological and especially geomorphological features that relate to the diversity of the ecosystem. The base of the region was formed approximately 200–145 Ma in the Jurassic; this layer is overlain by a granitic layer that was crated out of Cretaceous volcanism 98–72 Ma; the granitic base in turn is overlain by depositional layers from a shallow marine environment (the likes of tuff, mudstone, and sandstone) from 20 to 12 Ma, which were later pierced by intrusive rocks as magma upwelled again (Nemoto 2015; MOE-Tohoku nd). Crustal deformation and uplift occurred over the area through its entire geological history, and the heavy snowfall supplied water in the form of a potent erosive agent, leading to steeply incised hills where frequent landslides occur to this date. These dynamic earth features have sustained the beech forest and its biota over thousands of years. Due to the fact that the Siebold's beech

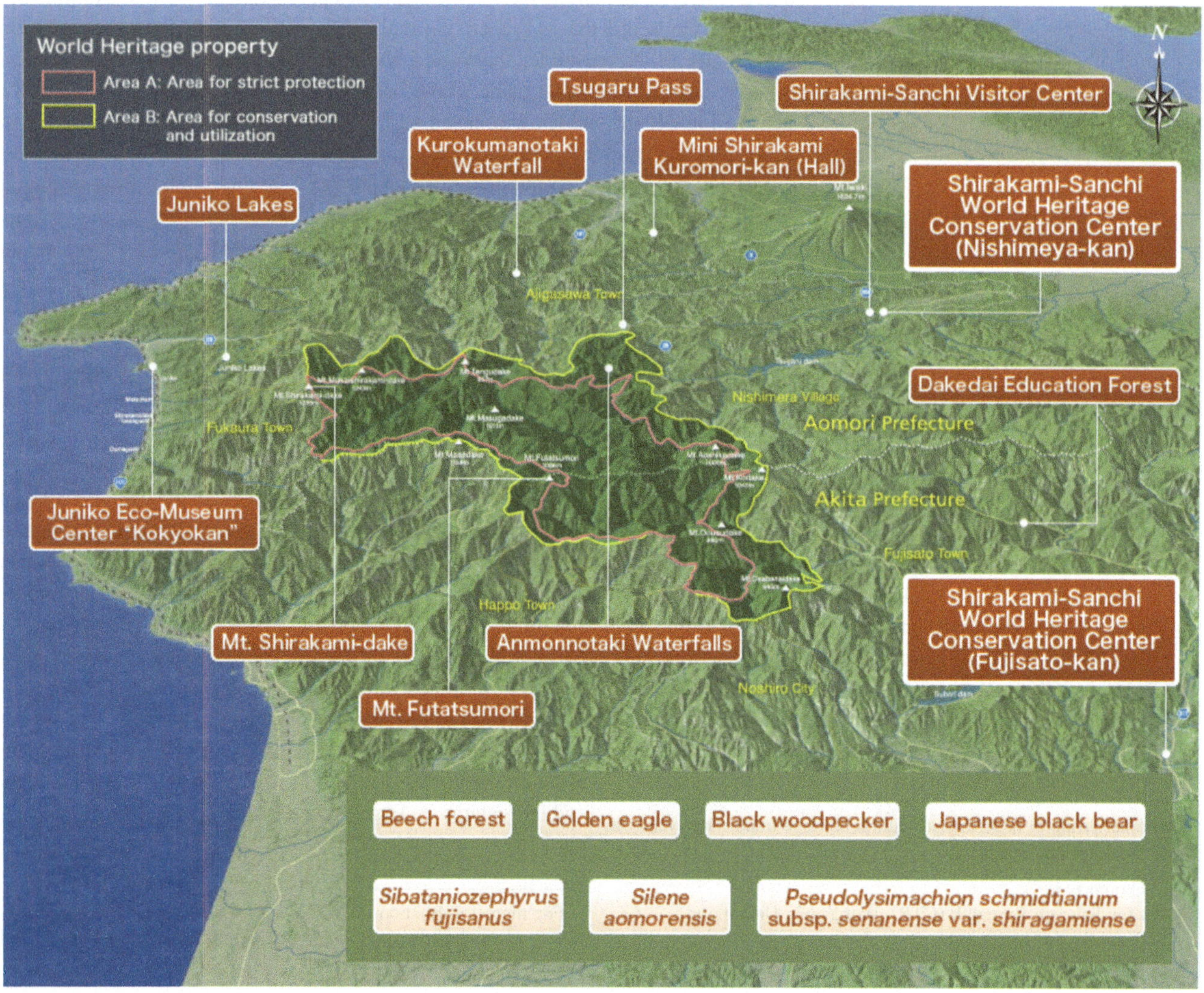

Fig. 5.1 Map of Shirakami-Sanchi World Heritage. Source: Ministry of the Environment website (World Natural Heritage in Japan) (https://www.env.go.jp/nature/isan/worldheritage/en/shirakami/area/index.html) (accessed on 2017/07/14)

forest has remained in a relatively stable condition for nearly 8000 years, it can be inferred that the landscape of Shirakami represents a fairly intact specimen of an ecosystem nearly spanning the entire Holocene in the Japanese Islands (Fig. 5.2).

5.2 Data Collection and Analysis Method

This chapter is based on an extensive analysis of literature (mostly available in Japanese). A field trip was conducted in 2015; the trip involved foot-transect method to observe the

Fig. 5.2 (**a**) A large specimen of *F. crenata* at the Takakura forest hiking trail (Photo copyright Makoto Nebuka, reproduced with permission). (**b**) Fall foliage in "Black woodpecker forest" (Kumagera no Mori) of Akaishi watershed (Photo copyright Makoto Nebuka, reproduced with permission)

landscape connectivity and fragmentation in the Akaishi River watershed, a visit to the World Heritage Center, and interviews with local resource persons. In keeping with the effort to *identify* patterns rather than bring out cause and effect relationships; research and analysis were based on a qualitative design.

5.3 Geological, Geomorphological, and Ecological Characteristics of the Area

The Shirakami Mountains span nearly 1300 km^2 (MOE nd), out of which 170 km^2 is designated as the World Heritage property (Fig. 5.3). The area at and around the Shirakami Mountains has a complex and composite geological structure. A detailed description is provided by Nemoto (2015), which can be summed up in the following manner. The oldest known base is formed of a mélange formation of chert, sandstone, and slate type rocks from the Jurassic; this formation hints at a submarine origin of this layer (Ibid, 20). The overlying layer of igneous rocks formed at a continental environment of the late Cretaceous as evidenced by the existence of mylonite in the rock layers; this layer shows extensive upwelling of magma and volcanism at that point (Ibid, 21). Early Miocene volcanism left its imprint in the form of the characteristic "green tuff" (formed out of ash and pumice). The upper part of this layer shows sandstone formation that hints at a shallow marine origin (ibid, 22). Layers above this (14–5 Ma) are dominated by mudstone formations, again providing evidence of a submarine environment. The entire region began to be uplifted rapidly from approximately 3.5 Ma; this uplift has created diverse coastal landforms such as marine terraces (common in the northwest of Shirakami Mountains and in the Happo Town of Akita Prefecture), and it also effects landslides that occur frequently in the mountains (Ibid, 25–26). The current rate of uplift is estimated as nearly 1.3 mm/year. (Higaki 2015). Although volcanism and deposition dominated the larger part of the long geological history, the current landforms and landscaping are dominated by uplift and erosion-driven processes. Higaki (2015) describes how the erosive processes themselves are complex (aided by the collapsibility of depositional layers, residual heat that alters rock properties (thermal weathering) leading to increased collapsibility) and how atmospheric and surface processes such as heavy snowfall, load of snow layer, freezing and thawing, and the potency of rivers all combine to aggressively erode surface features. These forces acting over time have sculpted the deeply incised mountains of Shirakami. The landscape here is thus in a state of perennial flux; this is a point that has obvious implications for management and conservation of this dynamic heritage.

Shirakami's topography is dominated by a chain of low to moderate elevation peaks, the highest point of which is Mt. Mukai-Shirakami (1243 m). Other notable peaks include Mt. Shirakami (1232 m), and Mt. Futatsumori (1086 m). Short and rapid rivers carrying heavy bedload incise this mountain chain and flow into the Sea of Japan. These rivers are also prone to frequent flooding. Main rivers of the area include Okawa, Akaishi, Oirase, Iwaki, and Sasanai in the Aomori Prefecture side and the Kasuke in the Akita Prefecture side (MOE nd). The Akaishi watershed is particularly notable for its geoecological characteristics; the watershed also came into focus during the 1980s when a plan to log

Fig. 5.3 A panoramic view of the Shirakami Mountains (Photo copyright Makoto Nebuka, reproduced with permission)

off what is now the central area of the World Heritage area emerged (Fig. 5.4).

Although an old-growth forest of large *Fagus crenata* forming the dominant canopy can lead to the perception that the forest environment is relatively unchanging over time, the ecological history of *F. crenata* is uniquely suited to change and fluctuation (Fig. 5.5). A detailed description of the forest composition of Shirakami can be found in Ishikawa (2015) and Nebuka (2011).

Ishikawa (2015) describes how the species dominates hillslopes that have shifted and collapsed repeatedly over time and details other localized colonies such as those of *Pterocarya rhoifolia*, *Pinus parviflora*, *Betula ermanii*, *Salix reinii*, and *Pinus pumila* (Fig. 5.6).

Fig. 5.4 *F. crenata* forest at the headwaters of Akaishi River (Photo copyright Makoto Nebuka, reproduced with permission)

Fig. 5.5 Hillslope movement and slope failure is a incessant landscape level process at Shirakami-Sanchi (Photo by author).

A total of 12 such varieties were reported by the comprehensive survey of the area's vegetation by Saito et al. (1995), who also report 21 varieties of shrub colonies including types that are adapted to riparian/wetland and rocky environments. In particular, *P. rhoifolia* is found in the stream corridors that are typically affected by flooding and bedload (silt, gravel) movement, and *P. parviflora* is adapted to the steep upper parts of hillslopes that are poor in water retention and prone to shifting (Ishikawa 2015).

Fig. 5.6 (**a**) Tight feedback between abiotic and biotic processes (such as vegetation pattern in areas affected by frequent landslides, and weathering of soil and rocks by the vegetation) makes these forests a valuable geoecological landscape (Photo by author). (**b**) The forest characteristics are strongly influenced by landscape processes and heavy snowfall, as evidenced by the appearance of large trees that experience these processes time and again photo by author)

In addition, during the foot transect of the Akaishi River I came upon several large and old specimens of *Aesculus turbinata* (Fig. 5.7) and *Magnolia obovata* that grow alongside the dominant *F. crenata* canopy (both species are common to hillslope areas). While it is difficult to census the species that specifically fall within the World Heritage boundaries, a total of 500 vascular plant species are reported (MOE nd), while the broader Shirakami Mountain area is known to house 1500 vascular plant species (Yamagishita et al. 2013). In addition, a variety of mushrooms are found in the forest; notable mushroom species are *Grifola frondosa*, *Mycoleptodonoides aitchisonii*, *Sarcomyxa serotina*, *Pholiota microspora* (Berk.) Sacc., *Armillaria mellea* subsp. *nipponica*), *Pleurotus ostreatus*, *Flammulina velutipes* (Curt.: Fr.) Sing., and *Lentinula edodes* (Yoshikawa 2015 personal communication; Nebuka 2011). Many of these species are adapted to broadleaf deciduous forests; mushrooms also constituted an important part of the diet of the local hunter-gatherer communities for many generations (Ibid). In addition, there are several species of edible herbs and plants that still form a part of the local traditional diet; notable varieties include *Chengiopanax sciadophylloides*, *Petasites japonicus* (Siebold et Zucc.) Maxim., *Elatostema umbellatum*, and *Osmunda japonica* Thunb. (Nebuka 2011).

As noted above, several animal and bird species find their habitat in the forest. The most notable large mammals are the Asiatic Black Bear (*U. thibetanus*), the serow (*C. crispus*), and the Japanese macaque (*M. fuscata*), while some important avian species are the Black Woodpecker (*D. martius*), the Mountain Hawk-Eagle (*N. nipalensis*), and the Japanese Golden Eagle (*A. chrysaetos japonica*). The Asiatic Black Bear is valuable both for its ecological role and as a symbolic species associated with at least one distinct type of hunting-gathering practice (the *matagi* community). The *matagi* hunting practice, conducted till the recent past, led to the formation of a valuable local ecological knowledge base. The short and rapid rivers crisscrossing the forest provide habitats for a number of fish species. Notable species include the salmonids (notably the Salvelinus char), *Plecoglossus altivelis*, *Oncorhynchus masou*, and *Cottus nozawae*. It is noteworthy that most varieties require a diversity of habitat conditions for spawning and growth: the fish are important markers of the continuity of an ecosystem that joins the Shirakami Mountains to the Sea of Japan.

The biodiversity of the forest allowed local communities to live off the forest for millennia. Widespread distribution of Neolithic ruins is reported from the area (Nebuka 2011), and until recently, the *matagi* community practiced a hunting-gathering lifestyle including hunting of the Asiatic Black Bear. Such practices engendered a rich traditional knowledge base of the forest which has been documented by Yamashita et al. (2013) and Nishiguchi (2006). Some scholars point out that the area is not a "pristine wilderness" in this sense (Nebuka 2011; Ishikawa 2015). While the history of settlement and the local traditions of living off the forest go back several millennia, it should be noted that anthropogenic

Fig. 5.7 A particularly large specimen of *Aesculus turbinata* at Akaishi watershed (Photo by author)

impact on the forest and species habitat has completely changed in nature from the Neolithic era to the hypermobile and industrialized lifestyle of today and that even communities who live within the Shirakami area no longer practice a strictly hunting-gathering lifestyle. This implies that any claim to *continuity* of the foraging traditions from several thousand years ago should be critically evaluated alongside the current level of anthropogenic stress on the landscape and the fragmented reality of the forest ecosystem. The issue of fragmentation is further detailed below.

Heritage in a Nutshell: Shirakami Mountains
Location: Aomori and Akita Prefectures in northern Honshu, Japan
Type of Heritage: Composite Geoecological (ecosystem of exceptional diversity and value underpinned by geological and geomorphic processes).
The property is highly evaluated for Exceptional Ecological Value of the old-growth Siebold's beech (*Fagus crenata*) forest and is a rare example of an old-growth natural forest in mainland Japan that retains a comparatively high degree of integrity of its natural environment, especially at the core of the World Heritage area.
Shirakami Mountains (literally "mountains of white gods") are a series of low to moderately elevated mountains (maximum elevation of 1243 m asl. is found at Mt. Mukai-Shirakami). The mountains are deeply incised by rivers that flow to the Sea of Japan. Apart from the Siebold's beech (very large specimens of which are found in all over the area), a number of broadleaf varieties and locally adapted vegetation colonies are found in this area, and the larger Shirakami Mountains area is known to house nearly 1500 types of vascular plants. The forests are also home to the Asiatic Black Bear and the Serow and many species of smaller mammals and birds.
Access: The core area can be accessed on foot from nearest trailheads, but the forest of the core area is not suited for casual walkers, and should only be accessed with relevant knowledge and preferably with experienced guides. Major access points are via Hirosaki City and Ajigasawa Town in the Aomori Prefecture side and along the JR Gono Line along the Sea of Japan in Akita Prefecture.
What to See: A variety of ecotourism activities are available in the area, including hiking, river trekking, and canoeing. However, heavy snowfall renders most trails inaccessible from winter to early spring. The World Heritage Visitor Center of the Ministry of the Environment is an excellent facility where visitors can learn about the natural heritage and conservation history of the forest.
Useful websites:
MOE Shirakami-Sanchi https://www.env.go.jp/nature/isan/worldheritage/en/shirakami/index.html
Shirakami-Sanchi Visitor Center English homepage http://www.experience-shirakami.com/en/

5.4 Logging Plan and Conservation Movement

The biggest threat to the old-growth forest of the Shirakami Mountains appeared in the late 1970–1980s, when a plan to log large sections of the forest and the construction of a "forest road" for timber extraction surfaced. This development led to local protest movements and involvement of nature conservation professionals who called for preserving the forest, eventually culminating in the stoppage of the planned forest road. The issue involved multiple stakeholders and viewpoints; a brief outline is provided below, following Sato (2006)'s documentation.

The proposal to construct the Aomori-Akita Forest Road (abbreviated Seishu Forest Road) was mooted in 1978, and the plan was adopted in 1980 by the forestry agency. The aim was to log large sections of the old-growth forest of Shirakami, procure forestry resources, and develop mines in the area (Ibid, Nebuka 1992). But the project was deeply unpopular as soon as it came to light, and it spurred nature conservation movements led by local citizens. In 1985, the original proposal was modified to change a section of the forest road from Fujisato Town in Akita Prefecture to the heart of the old-growth forest in the Akaishi watershed of Aomori Prefecture. Around this time (in 1986), national level nature conservancy groups such as the Nature Conservation Society of Japan (NACS-J) started advocating for a protected forest after the UNESCO MAB model. The development plan was supported by bureaucrats and administrative organs, and in the beginning protesters struggled to make their voices heard. However, local citizens, nature conservation enthusiasts, and some of the last *matagi* hunters came together against the developers and carried out a signature campaign. Extensive documentation was made of the ecological value of the forest, and the relationship between local society and the forest and threats from development were analyzed (Kamata 1987; Kudo 1987; Nozoe and Kitagawa 1995). This momentum eventually tipped the balance in favor of conservation, and the forest road plan was annulled in 1990, paving the way for conserving the area as a natural heritage.

While details of the citizen movement are outside the scope of this chapter, it is sufficient to observe that the last remaining old-growth beech forest of Japan and possibly the last large tract of wild forest in the Honshu Island was scheduled to be consumed by development pressure in the 1980s. Kamata (1987) provides details of fragmentation and partial logging that went on before the tide turned in favor of conservation. It can be concluded that the forestscape witnessed a significant phase of turmoil, recovery from which is likely to take several decades, if the forest and its adjoining areas can be kept relatively free from land conversion, urban and infrastructure development, and other anthropogenic pressures. This requires formulation and implementation of conservation plans as well as monitoring of the natural environment, preferably through local involvement. The task of effectively protecting the natural heritage of Shirakami is at once a sociopolitical and scientific challenge.

5.5 Fragmentation of the Forestscape and Issues Related to Managing Natural Heritage at Shirakami Mountains

As a World Heritage, the Shirakami Mountains are currently protected under several national and prefecture (local administration) level schemes. The MOE (nd) notes that currently the World heritage property is not under any significant threat of loss of its important characteristics. Major protection schemes include the Natural Environment Conservation Area (nearly 140 km^2 at the Core Area of the property, effective since 1992) and the National Wild Birds and Animals Preservation Area (nearly 170 km^2, effective since 2004) (MOE-Tohoku nd). In addition, the Shirakami-Sanchi World Heritage Area Scientific Council was established in 2010 in order to conduct scientific evaluation of conditions for preserving the heritage value (Ibid). While these schemes are robust and they currently protect the core region of the property effectively, issues relating to fragmentation in the wider landscape and potential threat from Global Environmental Change (GEC) remain. In the Akaishi watershed, a dam (Akaishi Dam) and a number of check weirs currently prohibit movement of fish and biomass through the river corridor (Fig. 5.8). Not only do dams inhibit the flow of the main channel, check (sabo) weirs have been constructed on numerous mountain streams that supply the Akaishi watershed with water, silt, and biomass. Although these alterations to the fluvial landscape occur outside of the core area of the World Heritage property, the ecosystem property and services of the watershed are nonetheless affected in a significant manner. In the Akaishi watershed alone, significant change in the transport of large bedload (pebbles, boulders) and visible biomass (woody debris) were reported (Yoshikawa 2015 personal communication, Nebuka 2015 personal communication). In addition, loss of older specimens of beech was reported (Nebuka 2015 personal communication). All of the reported changes hint at the fact that significant level of stress and fragmentation currently affects ecosystem resilience at the wider landscape level. As noted before, aggressive erosion, frequent landslides, and flooding are all major characteristics of the forest system and therefore occurrence of these events are vital signs of the ecological resilience of the forestscape. Anthropogenic alteration of surface level or fluvial features results in the loss of connectivity between different system components and a consequent erosion of ecological resilience; this must be effectively addressed at both planning and implementation levels. Additionally, catchments that have largely intact forest covers have a higher water retention capacity, which has potential implications for water management. This was evidenced by Kudo (2015)'s study which showed a catchment with a 90% beech forest cover (Anmon River) showing better water retention capacity compared to mixed forest catchments, at least for low- to moderate-level discharge events. Such flow regulation mechanisms are but one of the many important ecosystem services of the forest.

Since the World Heritage inscription, the Shirakami Mountains have witnessed a steady growth of tourism. Recently, a number of ecotourism schemes have taken root in the area (MoE nd); the area is also popular for forest trails, hiking mountains, river trekking, and sport fishing (Fig. 5.9). Shirakami Mountains were selected as one of the 13 "model ecotour areas" for 2004–2006 by the Ministry of the Environment (MOE), and in 2011, the Shirakami Ecotourism Promotion Council was established for planning and implementation of ecotours (MOE nd). An excellent description of the major trails in the region is provided by Nebuka (2011). Most trails can be walked with moderate level of preparation; however, due to heavy snowfall in the winter, most of these trails and their access routes are closed. The World Heritage Visitor Center of the MOE is an excellent facility equipped with scientific information, illustrative panels, and innovative displays that help the visitor understand the forest environment, as well as the history of nature conservation in the area.

Tourist access to the forest though could potentially lead to impact on the natural environment. Documentation of tourism impact on the forest landscape and ecosystems is

Fig. 5.8 Watershed level processes and connectivity are hampered by human intervention especially in the periphery of the World Heritage area. The photo shows a check weir on the Akaishi River (outside World Heritage area) (Photo by author)

scarce, but several local resource persons spoke of an apparent overemphasis on the Siebold's beech as an attraction but inadequate description of the connectivity of the forest ecosystems as far as tours are concerned. In at least one area of the Akaishi River watershed (outside the World Heritage zone), it was mentioned that a lack of knowledgeable stakeholders and the problem of fragmentation currently affects ecotour operability, although the area is popular for sport fishing of the *P. altivelis*. Thus, while the Shirakami Mountains remain an area with an excellent potential for wilderness education, ecotourism, and nature conservation thorough a participatory approach between managers and tourists, much of this potential is currently not realized. Efforts should be made to develop trails and tour packages in a way that incorporates visit to the World Heritage Center and participation in explanation and demonstration sessions.

Whereas the World Heritage inscription has helped preserve the forest of Shirakami—the core of which, at least, remains comparatively free from fragmentation—the dissociation of the local hunter-gatherers from the forest is an oft-mentioned issue. Especially, the relationship between the *matagi* community and the forest has been highlighted in this regard (Sato 2006). This community was a custodian of the forest and possessed rich traditional knowledge of the ecosystem. Sato observes that the restriction of access policy has been detrimental for managing the area as the *matagi* hunters are effectively barred from leading their traditional lifestyle. However, the current condition of this community and the feasibility of inter-generational continuity of the hunter-gatherer lifestyle have not been explored in comprehensive detail. The *matagi* community is currently extensively fragmented; younger generations have moved away from the hunter-gatherer lifestyle. While the concept of a self-sustaining forest community acting as custodians of the ecosystem and stewarding a World Heritage destination is appealing; the feasibility of such local stewardship needs to be assessed not by reminiscing the past but through rigorous analysis of intra- and inter-generational capacity of the local society.

a

b

Fig. 5.9 (**a**) Kurokuma Waterfall can be accessed from Ajigasawa town (Photo by author). (**b**) Aoike pond at Juniko Lake area in Akita Prefecture; it is one of the most popular tourist destinations of the area (Photo copyright Makoto Nebuka, reproduced with permission)

5.6 Conclusion

This chapter analyzed the Shirakami Mountains as a complex and dynamic natural heritage that involves interaction between distinct geological, geomorphological, and ecological features. The geological history of the area dates back to the Mesozoic, but the topography of the mountains and forests we see today is dominated by late Neogene and Quaternary uplift and erosion. It was shown how this dynamic environment is shaped by the interactions of uplift, erosion, and atmospheric processes; and how those intercations create a complex ecosystem that is dominated by *F. crenata* but is also shared by a large number of flora and fauna. The forest of *F. crenata* here is the last remaining wild Siebold's beech forest in Japan; it is an immensely valuable relic in the sense of time and is also a thriving ecosystem shaped by a species that reflects the ecological history of the Holocene. Although currently protected as a World Heritage property, the forest came to the brink of serious exploitation in the 1980s when a forest road to log off and transport forest resources was proposed. A successful protest movement by local citizens led to documentation and raised awareness of the unique ecological value of the forest landscape. However, although the core of the World Heritage is protected, significant fragmentation has occurred over the larger landscape; deforestation, land conversion, and damming of rivers outside the World Heritage area are some notable examples. Such changes could lower the ecological resilience of the forest and GEC could add further stress to the system. In order to effectively protect the forest of Shirakami, attention needs to be paid to its geological, geomorphological, and ecological connectivities, and heritage management plans that involve both scientific appraisal and traditional ecological knowledge should be implemented.

Acknowledgement I thank Makoto Nebuka for kindly providing some of the photographs.

References

Higaki D (2015) Tochi to Shinrin no Musubitsuki wo miru: Shirakami Sanchi no Jisuberi (Exploring the connections between soil and forest: landslides in the Shirakami Mountains). In: The Shirakam i Institute for Environmental Sciences, Hirosaki University (ed) Shirakamigaku Nyumon (Introduction to the study of Shirakami Mountains). Hirosaki University Publication, Hirosaki, pp 27–36

Ishikawa Y (2015) Sekaiisan to shite Shirakami seitaikei to Shokubutsu gunraku (Shirakami ecosystems and plant colonies as a World heritage). In: The Shirakami Institute for Environmental Sciences, Hirosaki University (ed) shirakamigaku nyumon (Introduction to the study of Shirakami Mountains). Hirosaki University Publication, Hirosaki pp 1–10

IUCN (nd) Protected areas categories. https://www.iucn.org/theme/protected-areas/about/protected-areas-categories. Accessed 12 Apr 2017

Kamata K (1987) Shirakami Sanchi ni Ikiru (Living in the Shirakami Mountains). Hakusuisha, Tokyo

Kudo S (1987) Shirakami Sanchi no Shokubutsu (Vegetation of Shirakami Mountains). Akita Sakigake Shinpo, Akita

Kudo A (2015) Shirakami Sanchi wa Midori no Damu ni Narieruka: Shirakami Sanchi ga Karyu Kasen ni Oyobosu Eikyo (Can Shirakami Mountains act as a green dam?: The effect of Shirakami forests on downstream drainage). In: The Shirakami Institute for Environmental Sciences, Hirosaki University (ed) Shirakamigaku Nyumon (Introduction to the study of Shirakami Mountains). Hirosaki University Publication, Hirosaki, pp 60–65

MOE (nd) Shirakami Sanchi. https://www.env.go.jp/nature/isan/worldheritage/shirakami/index.html. Accessed 12 Apr 2017

MOE-Tohoku (nd) The Shirakami Sanchi World Heritage Conservation Center. http://tohoku.env.go.jp/nature/shirakami/. Accessed 12 Apr 2017

Nebuka M (1992) Mori wo Kangaeru: Shirakami Buna Genseirin kara no Hokoku (Thinking of forests: report from the old-growth beech forestland of Shirakami). Rippu Shobo, Tokyo

Nebuka M (2011) Sekaiisan Shirakami Sanchi: Shizen taiken, Shisatsu, Kanko Gaido (World Heritage Shirakami Mountains: guidebook for nature experience, field observation and tourism). Nanatsumori Shuppan, Tokyo

Nebuka M (2015) Personal interview with author

Nemoto N (2015) Shirakami no Daichi no Oitachi (Geological profile of Shirakami Mountains). In: The Shirakami Institute for Environmental Sciences, Hirosaki University (ed) Shirakamigaku Nyumon (Introduction to the study of Shirakami Mountains). Hirosaki University Publication, Hirosaki, pp 20–26

Nishiguchi S (2006) Shirakami Monogatari: Matagi ga Aishita Kiseki no Yamayama (Shirakami chronicles: mountains of miracle loved by the Matagi). Shinpusha, Tokyo

Nozoe K, Kitagawa T (1995) Sekaiisan Shirrakami Sanchi kara no Hasshin (Report from Shirakami Mountains World heritage area). Doyukan, Tokyo

Saito N, Saito M, Makita H, Naito T (1995) Shirakami Sanchi no Shokusei (Flora of Shirakami Mountains). Shirakami Sanchi Shizen Kankyo Hozen Chiiki Sogo Chosa Hokokusho (Comprehensive report of the Nature conservation area of Shirakami Mountains). Kokuritsu Koen Kyokai, Tokyo

Sato M (2006) Shirakami Sanchi: Mori wa Yomigaeruka (Shirakami Mountains: will the forest be reborn?). Ryokufu Shuppan, Tokyo

UNESCO (nd) Shirakami Sanchi. http://whc.unesco.org/en/list/663. Accessed 12 Apr 2017

Yamashita Y, Nebuka M, Moriyama Y (2013) Shirakami Gaku: Shirakami Sanchi no Matagi (Shirakami studies: The Matagi of Shirakami Mountains). Buna-no-sato Shirakami, Nishimeya

Yoshikawa T (2015) Personal interview with author

6 Ogasawara Islands World Heritage Area: An Outstanding Ecological Heritage

Masahito Yoshida

Abstract

The Ogasawara Islands are created in the process of island-arc formation due to the ongoing subduction of an oceanic plate; these islands have witnessed unique speciation histories due to their remoteness from mainland Japan and any other continental landmass. The Ogasawara Islands were inscribed onto the UNESCO World Heritage List as a Natural Property due to high species endemism, adaptive radiation, and low extinction rates, but the total landform–landscape–ecosystem uniqueness and value deserves to be upheld as a natural heritage. Today, these islands have become one of the premier ecotourism destinations in Japan, but at the same time, tourism has the potential to negatively affect these isolated environments. This chapter explains the outstanding universal value of this heritage and analyzes the threat of invasive species for native ecosystems and challenges for managing tourism in a way that can help preserve this highly valuable system.

Keywords

Ogasawara Islands • Natural Heritage • Endemism • Adaptive radiation • Invasive species • Development impact

6.1 Introduction: Ogasawara Islands: A Unique Ecological Landscape and its Geological and Geomorphological Background

The Ogasawara (or Bonin) Islands are a group of volcanic islands located in the Pacific Ocean nearly 1000 km from Tokyo (World Heritage Committee 2011). Consisting of more than 30 islands, the Ogasawara Islands extend 400 km from south to north. These islands are divided into four island groups: namely, Mukojima Islands Group, Chichijima Islands Group, Hahajima Islands Group, Kazan Islands Group, and Nishinoshima Island (an active volcano) (Fig. 6.1). Among them, only Chichijima Island and Hahajima Island are currently inhabited. The World Heritage area covers 7939 ha including a terrestrial area of 6358 ha and a marine area of 1581 ha (Government of Japan 2010).

Geologically these islands are a part of the Izu-Ogasawara Arc, which is an excellent example of ongoing island arc formation due to the subduction of the Pacific Plate. When this remote island group was inscribed onto the UNESCO World Heritage (Natural Property) list, the main emphasis of its value rested on the unique biota. The islands were taken as an extremely valuable example of evolution of life in small oceanic islands—especially the concept of "adaptive radiation" where initially identical species undergo distinctive evolution due to distance from their original location and environmental constraints (for adaptive radiation, see Schluter 2000). Exceptionally high levels of endemism in vascular plants (of which around 440 species inhabit the islands) and land snails were mentioned as another main reason for inscription. In addition, low

M. Yoshida (✉)
Graduate School of Comprehensive Human Sciences, University of Tsukuba, Tsukuba, Japan
e-mail: yoshida_masahito@heritage.tsukuba.ac.jp

A. Chakraborty et al. (eds.), *Natural Heritage of Japan*, Geoheritage, Geoparks and Geotourism,
DOI 10.1007/978-3-319-61896-8_6

Fig. 6.1 Location of Ogasawara Islands

extinction rate (signifying a comparatively undisturbed ecosystem) and the combination of high species endemism, the range of adaptive radiation, and evolutionary variance between and within the islands are all recognized as important aspects of the property (World Heritage Committee 2011). As mentioned above, the islands are a product of oceanic plate subduction and submarine volcanism; landform evolution due to volcanism and aggressive erosion in the subtropical environment lead to landforms such as karsts. A rare example of undersea karstic landscape can be found near the Minamijima Island (Government of Japan 2010). All these aspects are interrelated and the islands therefore provide an excellent example of geological, landform, and biotic heritage that needs protection and sustainable management.

Ogasawara Islands are managed today as one of the leading ecotourism destinations in Japan. Due to the remote location, low level of human footprint on the landscape, and awareness of possible damage to the natural environment from unplanned visitations shared in the scientific and local communities, the area retains a naturally rich environment. Tourists enjoy activities such as trekking and touring in the islands as well as whale watching, diving, and kayaking off the islands. Though the tourism activities are conducted in a responsible manner, there is a possibility of unseen impact, and this necessitates continuous and stringent monitoring.

6.2 Outstanding Universal Value of Ogasawara Islands

The Ogasawara Islands were inscribed onto the World Heritage List in 2011 for significant ongoing ecological and biological processes in the evolution of terrestrial, freshwater, coastal, and marine ecosystems and communities of plants and animals (Criterion ix). These oceanic islands have never been connected to any continental landmass. Many oceanic islands as such are inscribed on the World Heritage List, but the Ogasawara islands have unique geological, geomorphological, and ecological characteristics. While the Hawai'i and Galapagos Islands are volcanoes that erupted from "hot spots" of the Pacific Plate,[1] Ogasawara are an oceanic island arc created by subduction of the Pacific Plate beneath the Philippine Sea Plate. It is postulated that this subduction began 48 Ma (Government of

[1] "Hot spots" are postulated as locations directly above mantle plumes that create active volcanoes when unusually hot magma pierces the Lithosphere (Morgan 1971). However, the hypothesis has courted controversy since it was first proposed to explain the formation of the Hawaiian Islands, and the a priori existence of superhot mantle plumes is an increasingly criticized concept now. For more details, see Foulger (2010).

Japan 2010). Ogasawara Islands are the only place in the world where we can witness the ongoing geological process of creation of oceanic island arc on land. Therefore, the Ogasawara Islands provide an excellent example of the ongoing process of island arc genesis and evolution, which are made into a dynamic geo-biospheric heritage for their unique ecological evolution.

Ogasawara Islands are located at subtropical zone (24°to 27°N in latitude). Because of the prevalence of the North Pacific High, the islands receive low precipitation in general. Mean annual precipitation of 1276 mm in Chichijima Island group creates conditions ideal for sclerophyllous scrubs (Fig. 6.2), while we can observe more precipitation at cloud forests in Hahajima Island (Government of Japan 2010). Sclerophyllous scrubs, mainly dominated by Shima-Isunoki (*Distylium lepidotum*) and Munin-Himetsubaki (*Schima mertensiana*), cover 477 ha (60%) of Anijima Island and 481 ha (20%) of Chichijima Island (Fig. 6.3). Subtropical rainforests, dominated by Ogasawara-Guwa (*Morus boninensis*) and Shima-Horutonoki (*Elaeocarpus phoniniaefolis*), grow up to 20 m height at the cloud forest belt of Hahajima Island (Fig. 6.4). Among the 441 species of vascular plants in Ogasawara Islands, 161 species are endemic. Woody plants have the higher endemic ratio of 64%; out of 138 woody plant species, 88 species are endemic (IUCN 2015).

Since Ogasawara Islands are remote, plants and animals that make up their ecosystems are descendants of pioneer species that arrived by floating on sea current, floating in the air or by flight. Due to this reason, the islands possess a unique terrestrial fauna that appear "unbalanced" from the point of usual terrestrial faunal evolution. For example, the Bonin Flying Fox (*Pteropus pselaphon*) is the only terrestrial mammal that inhabits Ogasawara Islands (it is listed as Critically Endangered in the Red List of Threatened Species by IUCN), while the Ogasawara snake-eyed skink (*Cryptoblepharus nigropunctatus*) and Gunther's tropical gecko or Micronesian gecko (*Perochirus ateles*) are the only two native reptile species that inhabit Ogasawara Islands (IUCN 2015). The islands do not have any amphibian species population.

Due to the advantage of wings, many birds have colonized the islands. Ogasawara Islands are home to many seabirds including the Short-tailed albatross (*Phoebastria albatrus*), Laysan albatross (*Phoebastria immutabilis*), Black-footed albatross (*Phoebastria nigripes*), and Red-footed booby (*Sula sula*). The Bonin honeyeater (*Apalopteron familiare*) is closely related to the Japanese white-eye (*Zosterops japonicus*) and the Golden white-eye (*Cleptornis marchei*) of Saipan, but the species has limited range and is endemic to Hahajima Island Group (Government of Japan 2010). The Ogasawara Islands are designated as Important Bird Areas (IBAs) and Endemic Bird Area (EBA) by Birdlife International.

Among invertebrates, insects (25% are endemic) and land snails (95% are endemic) show typically high endemism ratio in Ogasawara Islands (Government of Japan 2010). Endangered species belonging to the dragonfly and damselfly groups including the Ogasawara Tombo (*Hemicordulia ogasawarensis*) and the Ogasawara Aoitotombo (*Indolestes boninensis*) are found in these islands; these species are respectively evaluated as Critically Endangered and Endangered in the Red List of Threatened Species by IUCN (Government of Japan 2010; IUCN 2015).

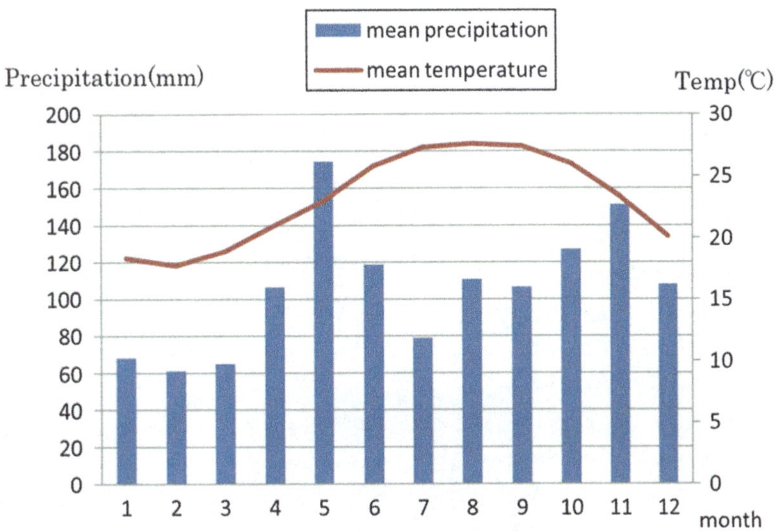

Fig. 6.2 Mean temperature and mean precipitation by month in Chichijima (Government of Japan 2010)

Fig. 6.3 Sclerophyllous scrub in Anijima Island

Fig. 6.4 Cloud forest in Hahajima Island

Fig. 6.5 Mandarina evolved in habitat of high trees

One of the most distinctive biological features of Ogasawara Islands is the adaptive radiation of land snails. Among the 104 species of land snails recorded in Ogasawara Islands, 98 (94%) (Government of Japan 2010) are endemic. Latest DNA studies of land snails show a distinctive history of adaptive radiation of genus Mandarina (Figs. 6.5 and 6.6) (Chiba 2004). In Ogasawara Islands, repeated speciation has occurred to adapt to various ecotypes from arboreal to ground within each island groups from Mukojima Islands Group, Chichijima Islands Group, and Hahajima Islands Group (World Heritage Committee 2011). Because these islands show valuable evidence of significant ongoing ecological processes including adaptive radiation in the evolution of land snails, they were inscribed on the UNESCO World Heritage List at the 35th session of World Heritage Committee in 2011.

6.3 World Heritage Inscription and Challenges

Although these remote oceanic islands possess unique geo-biological characteristics and constitute a globally significant natural heritage, the outstanding universal value of Ogasawara Islands has been threatened in recent years through the introduction of alien invasive species, the construction of an airport, installation of new passenger boats, and visits from foreign passenger ships; all of these may increase the impact on the natural environment of Ogasawara Islands in the near future. When Ogasawara Islands were inscribed on the World Heritage List in 2011, the World Heritage Committee requested the Government of Japan to continue its efforts to control invasive alien species and asked it to implement rigorous environmental impact assessment prior to any further significant infrastructure development (World Heritage Committee 2011).

6.3.1 Invasive Alien Species

Invasive alien species constitute a major threat to the biodiversity of Ogasawara Islands. Although eradication program of goats, pigs, cats, and rats have been successful in Mukojima and Chichijima Islands group, many issues concerning invasive alien species remain. Feral cats were great threats to seabirds and the wood pigeon (*Columba janthina nitens*; listed as Critically Endangered in the Red List of the Ministry of the Environment of Japan) (MoE 2012). Eradication of feral cats though was criticized by advocates of animal welfare, leading to a clash of interests. The Institute of Boninology, together with IUCN's Conservation Breeding Specialist Group of Species Survival Commission, held an International Workshop that was titled "Conservation Planning of Wood Pigeon" in Chichijima in 2008 (Horikoshi et al. 2010). This workshop played an important role to bring all stakeholders together for wood pigeon conservation as well as for addressing the issue of animal welfare (in the form of feral cats in this case). The feral cat eradication program has been conducted in a careful manner in Chichijima since 2008. Captive feral cats were not killed but were transferred to the clinic of Tokyo Veterinarian Association and they were later trained as companion animals. Population of wood pigeons has increased since 2011, which bears testimony to the success of this program (Sasaki 2014).

Fig. 6.6 Mandarina evolved in habitat under ground

Heritage in a Nutshell: Ogasawara Islands
Type of Heritage: Ecological, with geological and geomorphological underpinnings (oceanic island arc created by subduction providing uniquely valuable biotic evolution and speciation history).
Ogasawara are a group of 30 volcanic islands located 1000 km from Tokyo. These islands have never been connected to any continental landmass, resulting in their unique flora and fauna. The UNESCO World Heritage value is derived from high levels of endemism, low extinction rates, and adaptive radiation of species. Only Chichijima and Hahajima islands are currently inhabited.
Ogasawara Islands area leading ecotourism destination in Japan. Tourists can enjoy walking forest trails on the islands and enjoy marine animal watch (whales, dolphins), diving, and sea kayaking off the islands. Notable scenic points include Anijima Island Marine Park, Minamijima Island (karstic landscape), Minamizaki Beach, and Sekimon Forest. Lodging opportunities are limited to the Chichijima and Hahajima Island. The islands can only be accessed by marine transport (ships).
The area is also notable for information sharing about human impacts on the landscape and voluntary measures to control negative tourism impact. However, as these islands were never attached to any continental landmass, the environment is particularly vulnerable to alien species invasion.
Useful websites:
Ministry of the Environment Ogasawara National Park https://www.env.go.jp/en/nature/nps/park/ogasawara/guide/view.html
Ecotourism Ogasawara http://www.ogasawaramura.com/en/

The Green Anole (*Anolis Carolinensis*) is an alien lizard (originally native to North America) that was introduced through Guam Islands during the era of US occupation after the World War II (Fig. 6.7). Anoles eat up most of the insect population in Chichijima and Hahajima Island and were listed as *Designated Invasive Alien Species* under the Invasive Alien Species Act in 2005. The distribution of Green Anole was initially limited to human-inhabited islands, but a small population was found in 2013 in Anijima Island which is home for many endemic insect species. The Scientific Council of Ogasawara World Heritage, consisting of 12 scientists, issued an emergency call after the discovery of Green Anole in Anijima. The Ministry of the Environment together with the Forestry Agency of the Japanese national government and the Tokyo Metropolitan Government launched a program to eradicate the established population and to prevent further expansion of Green Anole in Anijima. The main response was to set up a long fence that divided Anijima from east to west that was supposed to inhibit green anole movement across the island (Fig. 6.8) (Ogasawara World Heritage Area Regional Liaison Committee 2013). Despite these efforts, the anole population of the island has not been eradicated.

The New Guinea Flatworm (*Platydemus manokwari*) is a carnivorous flatworm; it was introduced as a predator of East

Fig. 6.7 Green Anole in Chichijima

Fig. 6.8 Fence constructed in Anijima to prevent expansion of Green Anoles

African Land Snail (*Achatina fulica*), another invasive species. While the population of the East African Land Snail has decreased, the New Guinea Flatworm itself became a big threat to endemic land snails of Ogasawara Islands. The flatworm was listed as a *Designated Invasive Alien Species* in the Invasive Alien Species Act in 2005. Since the distribution of New Guinea Flatworm is restricted to Chichijima Island, the Ministry of the Environment, the Forestry Agency, and the Tokyo Metropolitan Government launched a program to prevent further expansion to important habitats of endemic land snails in Chichijima and prevent flatworm invasion of other islands. Tourists are encouraged to clean their shoes with brushes and spray vinegar onto the soles before entering important habitats of endemic species in Chichijima, or before entering islands where flatworm populations are currently not detected. The vinegar is used because it was found that flatworms are vulnerable to vinegar (Ministry of the Environment 2015) (Fig. 6.9). The Ministry of the Environment set up an electric fence to prevent expansion of flatworms into the endemic land snail habitats since 2014. Further efforts are required to clean up the equipment, construction vehicles, and scaffolding pipes that have been moved from Chihijima to other islands. Introduction of mango or pineapple saplings from Okinawa to Hahajima also constitutes a potential threat to endemic land snails, because soil in the sapling pot may include carnivorous flatworms and other invertebrates. Alien ants, diplopods, and carnivorous flatworms were found from the soil in the pods of Mango saplings introduced from Okinawa (Working Group on Prevention of Introduction of New Alien Species 2016). To prevent new invasive alien species, introduction of soil cleaning measures using hot water is effective (Fig. 6.10).

The Big-headed ant (*Pheidole megacephala*; believed to be native to southern Africa) is one of the invasive ant species that possesses a significant threat along with the Argentine ant (*Linepithema humile*), Crazy ant (*Anoplolepis gracilipes*), Red imported fire ant (*Solenopsis invicta*), and Little fire ant (*Wasmannia auropunctata*). These species are listed on the *100 World's Worst Invasive Alien Species* compiled by IUCN (Lowe et al. 2000). The Big-headed ant has established its population at the Futami Port of Chichijima and was recently discovered in Hahajima in 2014. Recent study (Uchida et al. 2016) reveals a strong correlation between the decrease of the number of land snails and Big-headed ant populations in Hahajima[13]. Again, import of mango saplings from Okinawa as well as translocation of equipment and soil from Chichijima acts as possible transport pathways of these invasive ants. In a nutshell, the introduction of strict biosafety measures is essential to safeguard endemic land snails from invasive alien species.

Fig. 6.9 Shoes cleaning using brush and vinegar to prevent expansion of New Guinea Flatworm

Fig. 6.10 Soil cleaning using hot water to eradicate carnivorous flatworm

6.3.2 Transportation and Tourism Impact

Since becoming World Heritage Site, the Ogasawara Islands have seen a steady expansion of tourism-related activities. While most of the tourism activities are managed with environmental impact in mind and guidelines are in place to monitor tourist behavior, tourism expansion has a potential to affect the natural environment of these islands, which is not only precious but also fragile due to its isolation. Song and Kuwahara (2016) provide a description of how ecotourism came to be a viable management policy for the Ogasawara Islands, and they state that the Ogasawara Islands are an "advanced ecotourism area" in Japan. A variety of activities are available for the ecotourist such as nature/culture trails, sea kayaking, diving, snorkeling, and dolphin and whale watching. Song and Kuwahara also describe the emergence of whale watching as a form of ecotourism, noting how the Ogasawara Whale Watching Association was established in 1989, how self-imposed rules pertaining to whale watching were formulated in 1992, and how this association served the dual purpose of attracting ecotourists and conducting research. In 1996, the Nature Conservation Society of Japan (NACS-J) and the Ogasawara Village Administration joined hands to survey the tourism carrying capacity of Minamijima and formulate ecotourism use rules for that island. In 2000, the Ogasawara Tourism Association voluntarily adopted a set of rules to prevent overuse of the landscape (Ibid: 38). A training program of qualified ecotour guides has also been implemented in Ogasawara. However, some scholars have observed that there are several limitations in these efforts: Suzuki (2010) noted that the guide interpretation did not go beyond the obvious and most visible features (and species), and Ishihara et al. (2010) observed that the tourism management scheme in Ogasawara Islands is mostly done through top-down control by the Tokyo Metropolitan Government.

In the requests and recommendations section, the World Heritage Committee in 2011 had expressed its concern for new infrastructure facility development and impact of possible increase of tourist on Ogasawara Islands. IUCN particularly recognized the history of struggle for conservation since 1989, centering on the construction of an airport at Ogasawara Islands. The original airport construction plan developed by the Tokyo Metropolitan Government involved clear cutting of the Sclerophyllous scrubs at Anijima Island for runways and for constructing a ropeway to connect Anijima and Chichijima (Fig. 6.11) (Nature Conservation Society of Japan 2002). However, scientific studies revealed that Sclerophyllous scrubs at Anijima are important as the habitat of endemic land snails and insects (Chiba 1989). Opposed by scientists and conservation organizations, the Tokyo Metropolitan Government relocated the construction plan from Anijima to Chichijima. However, the proposed site again attracted controversy because it is the last remaining habitat of endemic plants such as *Rhododendron boninense*, a Critically Endangered species on the Red List of the Ministry of the Environment. Eventually, the Tokyo Metropolitan Government gave up its plan to construct a commercial airport at Ogasawara Islands in 2001(Nature Conservation Society of Japan 2002). This turn of events favored the local ecosystems, but not all problems were solved. The Ogasawara Village did not give up its plan to construct a small airport for villagers, and it will be possibly detrimental to the environment in the light of the fact that the World Heritage Committee had requested the Government

Fig. 6.11 Anijima where airport construction was planned in 1989

of Japan to undertake a rigorous environment impact assessment prior to the construction of infrastructure.

Prior to the inscription of Ogasawara Islands on the World Heritage List, the Tokyo Metropolitan Government signed contract with the Ogasawara Village concerning the appropriate utilization of Minamijima and the Sekimon Trail in Hahajima. Guides are obliged to attend a series of lecture on ecotourism and pass an examination. The number of visitors to designated areas must be carefully managed so that it stays within the carrying capacity of those areas (Fig. 6.12). The Ogasawara village has also launched its own training program for guides. Tourism in Ogasawara Islands is apparently well managed up to now, but we need to be cautious as the situation surrounding tourism in Ogasawara Islands is gradually changing. From March 2016, large passenger ships are allowed to anchor into the Futami Bay of Chichijima: this has the potential of further introduction of new alien species to the Ogasawara Islands. In July 2016, a new passenger boat Ogasawara-maru III came into service; this boat is larger and faster than its predecessor, Ogasawara-maru II, which carried tourists into the islands. These circumstances surrounding tourism can change the natural environment, and continuous monitoring of tourism and other development-related aspects is required.

6.4 Conclusion

This chapter reflected on how the unique geological setting (and geomorphological attributes) frames the outstanding ecological heritage of Ogasawara, for which it was given World Heritage (natural area) status. These remote islands offer a fascinating example of ongoing speciation and evolution of life, as well as dynamic earth processes. These islands are also among the leading ecotourism destinations in Japan, and they offer an exciting range of activities for the responsible ecotourist. However, there are several serious challenges such as invasive alien species that spread through

Fig. 6.12 Authorized tour guide leads tourists in Minamijima

human access and development plans. It was discussed how seemingly innocuous species such as the green anole, the New Guinea flatworm, and the big-headed ant can become serious threats to native species. It was also discussed how a planned airport facility could have seriously affected the endemic flora and fauna. While tourism in these islands is currently conducted with careful planning and consideration of impact on the environment, many of the possible threats are insufficiently understood and require continuous monitoring, especially because the environmental history of these islands is unique and is not found anywhere else.

References

Chiba S (1989) *Mandarina* in Anijima, Ogasawara Islands. Ogasawara Res 12:49–55

Chiba S (2004) Ecological and morphological patterns in communities of land snails of the genus *Mandarina* from the Bonin Islands. J Evol Biol 17:131–143

Foulger GR (2010) Plates vs plumes: a geological controversy. Wiley Blackwell, Chichester

Government of Japan (2010) Nomination of the Ogasawara Islands for inscription on the World Heritage List. http://ogasawara-info.jp/pdf/isan/recommendation_en.pdf. Accessed 6 May 2017

Horikoshi K, Itoh K, Hayama S, Murayama A, Suzuki H, Traylor-Holzer K, Ballou J, Miller P (eds) (2010) Red-headed wood pigeon population and habitat variety assessment workshop final report. Conservation Breeding Specialist Group, IUCN/SSC, Apple Valley, MN

Ishihara S, Kosaka W, Morimoto K (2010) Ogasawara Shoto no ekotsurizumu wo Meguru Chiiki Shakai no Shiko Sakugo: Nanto Ruru Kadai wo Chushin ni (Trial and error of regional society over the ecotourism of Ogasawara Islands: a matter of Nanto rule). Ogasawara Kenkyu Nenpo 33:7–25 (In Japanese).

IUCN (2015) The IUCN Red List of Threatened Species. http://www.iucnredlist.org. Accessed 6 May 2017

Lowe S, Browne M, Boudjelas S, De Pooter M (2000) 100 of the World's worst invasive alien species: a selection from the global invasive species database. Invasive Species Specialist Group, SSC, IUCN

Ministry of the Environment (2012) Red list of threatened species, 4th edn. http://www.biodic.go.jp/rdb/rdb_f.html. Accessed 6 May 2017 (In Japanese)

Ministry of the Environment (2015) Ogasawara Shoto Gairai Planaria rui no Shinnyu Kakudai Boshi ni Kansuru Taiohoshin (Invasive alien species of Ogasawara—flatworms). Ministry of the Environment. http://ogasawara-info.jp/pdf/isan/guideline_invasive_1.pdf. Accessed 6 May 2017 (In Japanese)

Morgan WJ (1971) Convection plumes in the lower Mantle. Nature 230:42–43

Nature Conservation Society of Japan (2002) Shizenhogo NGO Hanseiki no Ayumi (Fifty years record of nature conservation NGO). Heibonsha, Tokyo (In Japanese)

Ogasawara World Heritage Area Regional Liason Committee (2013) Ogasawara Shoto Anijima ni okeru Green Anole Tanki Bojo Keikaku (Short term eradication plan of Green Anole at Anijima). http://ogasawara-info.jp/pdf/chiiki2501/2501_shiryou1-2.pdf. Accessed 6 May 2017 (In Japanese)

Sasaki T (2014) Ogasawara Shoto no Neko Taisaku- Genzai no Stage (Cat control in Ogasawara Islands—current stage). Toju J 564:18–19. http://tvma.or.jp/news/items/img-516134839.pdf. Accessed 6 May 2017 (In Japanese)

Schluter D (2000) The ecology of adaptive radiation. Oxford University Press, Oxford

Song D, Kuwahara A (2016) Ecotourism and World heritage: its influence on islands in Japan. J Mar Island Cult 5:36–46

Suzuki K (2010) Poritikusu to shite no Sekai Isan (World heritage as politics). Kanko Kagaku Kenkyu 3:57–69 (In Japanese)

Uchida S, Mori H, Kojima T, Hayama K, Sakairi Y, Chiba S (2016) Population decrease of land snails by introduction of Big-headed ant (*Pheidole megacephala*) into Ogasawara Islands. ESJ63 P1-389. Ecological Society of Japan

Working Group on Prevention of Introduction of New Alien Species (2016) Sekai Isan Chiiki Ogasawara Shoto Aratana Giraishu no Shinnyu Boshi ni Kansuru Kento no Seika to Kongo no Kadai no Seiri (Strategy to prevent introduction and expansion of alien species in Ogasawara Islands World natural heritage area). http://ogasawara-info.jp/pdf/isan/challenge_invasive%20alien%20species_3.pdf. Accessed 6 May 2017 (In Japanese)

World Heritage Committee (2011) Decision 35 COM 8B.11 Ogasawara Islands (Japan). http://whc.unesco.org/en/decisions/4282. Accessed 6 May 2017

Yakushima Island: Landscape History, World Heritage Designation, and Conservation Status for Local Society

7

Shigemitsu Shibasaki

Abstract

This chapter provides a description of Yakushima Island, off the coast of Kyushu in southwestern Japan, which was designated as a World Heritage Site (Natural) in December 1993. Yakushima has an image of "pristine nature," and the island contains several rare and relatively undisturbed geological, geomorphological, and ecological features. This perception attracts many tourists to the island, but conservation areas of Japan also contain a wealth of cultural and historical materials such as communal forestry and sacred landscapes, in addition to their natural landscapes. This chapter presents an overview of the human association on the forests in modern times and provides an account of some challenges that world heritage designation can pose for the management of heritage. The chapter also focuses on how cultural and historical remains and the customs of local communities constitute an important addition to the recognized natural heritage values.

Keywords

World Heritage Sites • Cultural activity • Pristine nature • Heritage management

7.1 Yakushima: An Outline of the World Heritage Property

Yakushima is an island administered by Yakushima-cho Town as a part of Kagoshima Prefecture. It is located approximately 70 km offshore from Cape Sata,[1] and approximately 135 km to the south of Kagoshima Port (Fig. 7.1). Yakushima is connected to the main islands of Japan through Yakushima Airport, Miyanoura Port, and Anbo Port. The island can be visited by air, ship, and hovercraft. The area of the island is 505 km^2 and its diameter is around 25 km.

Yakushima's unique island ecosystem is characterized by distinct vertical distribution of vegetation as well as its peculiar forest landscapes such as natural forests of *Cryptomeria japonica* (Sugi or Japanese cedars) (IUCN 1993). Highly evaluated under Criteria IX and VII, respectively, an area of nearly 107 km^2 (10,747 ha or about 21% of Yakushima Island) was inscribed onto the UNESCO's World Heritage List in December 1993 as one of the two first Natural Heritage Properties from Japan.[2] Approximately 95% of the designated area is national forests managed by Forestry Agency, (organ of the Ministry of Agriculture, Forestry, and Fisheries) while the Kagoshima Prefecture owns the rest. The designated area belongs either to the Yakushima National Park or to the Yakushima Wilderness Area; both of which are managed by the Ministry of the Environment (MoE). Yaku-sugi forests are also designated as Special Natural Monuments administrated by the Agency for Cultural Affairs (affiliated to the Ministry of Education, Culture, Sports, Science, and Technology). The current Yakushima World Heritage Area Management Plan was collaboratively

[1] The southernmost tip of Kyushu Island.

[2] Shirakami-Sanchi (Aomori prefecture and Akita Prefecture) is the other property listed in the same year.

S. Shibasaki (✉)
The National Museum of Japanese History (REKIHAKU), Sakura, Japan
e-mail: shiba@rekihaku.ac.jp

A. Chakraborty et al. (eds.), *Natural Heritage of Japan*, Geoheritage, Geoparks and Geotourism,
DOI 10.1007/978-3-319-61896-8_7

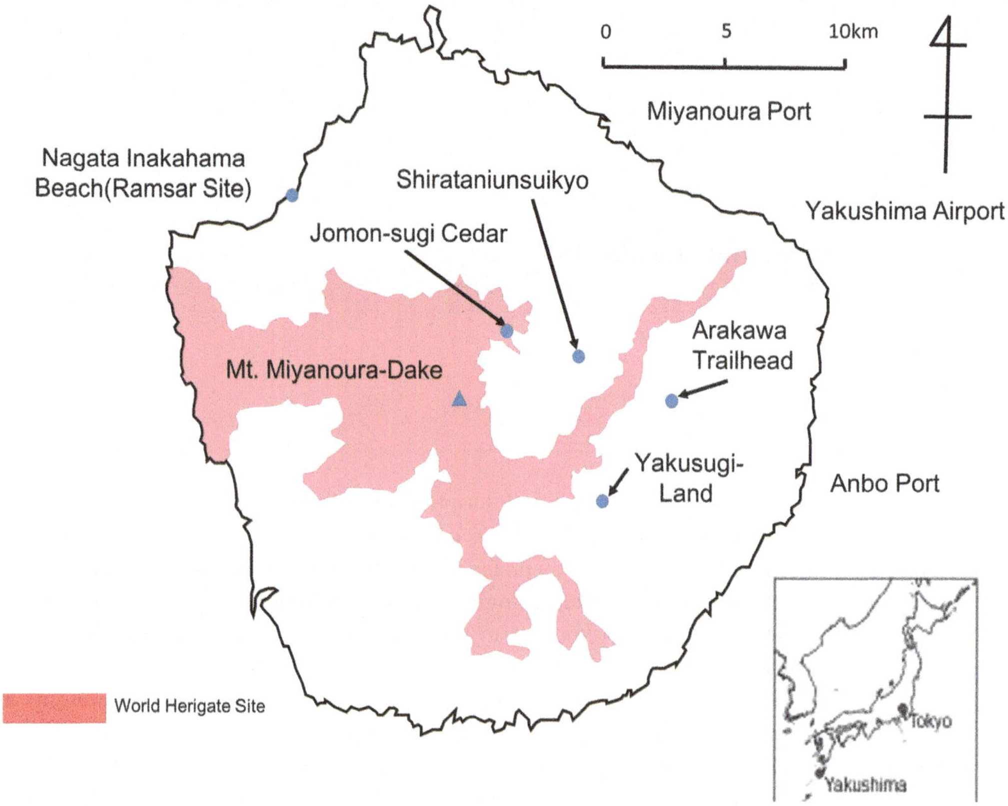

Fig. 7.1 Yakushima Island. Source: Modified and updated version of Shibasaki et al. (2006)

prepared by MoE, the Forestry Agency (FA), the Agency for Cultural Affairs, Kagoshima Prefecture, and Yakushima-cho Town in 2012 (MoE et al. 2012). This system allows for combined and holistic management of both natural and cultural interpretations of the environment of the island.

7.2 Geology, Flora, and Fauna

Yakushima is formed by a combination of sedimentary and orogenic processes (Sato and Nagashima 1979). The base rock—the Kumage group—is an accretionary complex that accreted to the Eurasian Continent in Eocene (ca.40 Ma) and is composed of sandstone and mudstone that deposited on the ocean floor and along the trench (Saito et al. 2007). Around 15.5 Ma, granite began to intrude into the Kumage group stratum, and the subterranean granite rose up to almost 2000 m asl in 15 million years, eventually creating the dramatic mountain landscape that exists today (Saito et al. 2008). Yakushima has no volcanoes, but a volcanic ash layer deposited by the eruption of Kikai caldera located some 40 km northwest of Yakushima about 7300 years ago, is widely distributed on the island (MoE et al. 2012). The pyroclastic flow caused by this eruption reached Yakushima Island (Koya Pyroclastic flow deposit), and this event had a disastrous impact on local culture and the natural environment (Yakushima Environmental Culture Foundation 2000). These orogenic processes created Yakushima's steep topography. The mountains standing right up to the seashore on Yakushima are locally known as "Mae-dake," meaning "foreground peak," while larger peaks in the central area are called "Oku-dake" or inland peaks. The latter include Mt. Miyanoura-dake (1936 m asl) the highest peak in the Kyushu area.[3] The island's topography and its location also influence its climate. Warm moist air rising out of the Kuroshio Current collides with the seaside cliffs and forms rainclouds. Yakushima is so rainy that it has metaphorically expressed that "Yakushima has 35 rainy days in a month" (Hayashi 1951, 416). The highest rainfall locations have recorded annual precipitation of 10 m, and the upper elevations of mountains such as Mt. Miyanoura-dake are covered by several meters of snow in winter.

These conditions, along with the fact that Yakushima is a small island separated from the larger, more heavily impacted natural environment of Kyushu, have resulted in

[3] Although situated away from the main island of Kyushu Yakushima falls under the Kagoshima Prefecture and therefore its mountains are commonly referred to as in Kyushu area.

a rich flora. A total of over 1900 species and subspecies of flora are noted, and around 600 species of moss and 230 species of plants find their southern habitat limit on this island (MoE et al. 2012). Yakushima's flora contains 41 critically endangered species, 52 endangered species, 73 vulnerable species, and 31 near-threatened species based on the National Red List Category (MoE 2007); and these include 94 species endemic to the island, including subspecies and varieties (MoE et al. 2012).

Cryptomeria japonica (Sugi or Japanese cedar) is a part of the natural vegetation in middle parts of the mountains (about 600–1800 m asl) (Osawa 2006). Some of the specimens have lived over 700–800 years and have attained gigantic proportions (Kagoshima Dairinkusho 1923). These old Japanese cedars are called Yaku-sugi[4] and form an important tourist attraction. The "Jomon-sugi," with the thickest breast high diameter, attract travelers who trek for around 10 h (round-trip time) to see it. Elsewhere in the Maedake range (0–800 m asl.), laurel-leaved forests include *Quercus serrata* Murray (Konara), *Castanopsis sieboldii* (Sudajii), and *Distylium racemosum* (Isunoki). Coastal locations feature a significantly different vegetation of subtropical nature, such as *Ficus microcarpa L.f.* (Gajumaru), *Ficus superba* var. *japonica* (Akou) (Yakushima Environmental Culture Foundation 2000). Also, there is a small mangrove forest composed of *Kandelia obovata* Sheue (Mehirugi) along the south-western coast. These forest resources have been used by island communities for many years; for example, local people used to cut laurel-leaved trees and produce charcoal (Inamoto 2006).

Yakushima Island is also home to a diverse range of fauna. There are 16 species of mammals including 4 subspecies such as the *Macaca fuscata yakui* (Yaku-zaru macaques) and *Cervus nippon yakushimae* (Yaku-shika deer) that are endemic to the island; furthermore, 167 bird species are found on the island, mainly in the coastal lowlands (MoE et al. 2012). Yakushima is a stopping point for many migratory species. In addition, there are 15 species of reptiles, including the *Caretta caretta* (Loggerhead turtle) and *Chelonia mydas* (Green turtle), 8 kinds of amphibians, and 1900 species of insects (Okano and Matsuda 2013). Yakushima Nagata-hama Beach, with the largest spawning frequencies for the loggerhead turtle in North Pacific Ocean, was designated as the Ramsar site in 2005 (Yakushima World Heritage Conservation Center nd).

7.3 Sociocultural Aspects of the Island

There are 24 communities in the coastal zone, and the size of the island's population was 13,378 as of May, 2016 (Yakushima-cho Town 2016). Farming of citrus fruits such as *Citrus reticulate* var. *poonensis* (Ponkan) and *Citrus tankan* Hayata (Tankan) by cultivating slopes, and fishing for flying fish and horse mackerel, used to be important sources of income for farmers on the island; however, these industries are affected by aging of the local society and a continuous drop in revenues in recent times. One of the alternative employment sources, the construction industry, matured dramatically at the end of the twentieth century due to subsidies for public services (especially road construction) on remote islands including Yakushima. At least 78 construction companies were operating in 2000; however, this industry has also shrunk in recent years (Shibasaki 2004).

At the end of 1980s, another alternative, the tourism industry started to develop. Just before the island's designation as a world heritage site, visitor centers and natural museums were constructed by public organizations (Shibasaki et al. 2006). During the peak season (from spring to summer), around 60% of tourists to Yakushima visit the Jomon-sugi in around 2010 (Shibasaki 2014a), and this involves a 10-hour-round trip between the Arakawa Trailhead and the Jomon-sugi cedar site.

7.3.1 Forestry on Yakushima

Although Yakushima is inscribed in the World Heritage List as a "natural" property, its rich landscape is far from pristinely natural. Here, the history of forestry at Yakushima can be recounted briefly to argue this point. In ancient times, the residents of the island respected and feared "Oku-dake" areas (the inner parts of the mountains) as the inner sanctum of gods, and rarely visited such sacred areas. However, that all changed during the Edo Period (1603–1868), when the local Satsuma Clan set its sights on the timber resources of the island, and asked for tax payment in terms of Sugi wood instead of rice (Yaku-cho Town Committee for Local History 2007).[5] At first, residents feared this decision because the ancient trees were worshipped as gods, but they relented after the mediation of a local priest Tomari Jochiku (1570–1655), who had been born in Yakushima's Anbo village, and worked at Satsuma Clan headquarters as a

[4] Literally "Yakushima cedar." Definitions of Yaku-sugi differ in various periods in history. Currently, it is common that Yaku-sugi should be aged over 1000 years, while ages of 700–800 years and over should be enough to be called Yaku-sugi cedars before World War II. In this chapter, the prewar definition will be used because history of forest policies before WWII is discussed.

[5] Some forest development was sporadically carried out in the area before the Edo period, but continuous forest development did not begin until the rule of the Satsuma Domain in Edo.

Confucian priest (Yakushima Environmental Culture Foundation 2000; Godai and Hashiguchi 1905).

From the middle of the seventeenth century, continuous forest exploitation began on the island (Yaku-cho Town Committee for Local History 2007). In those days, local people stayed in the mountainous areas for up to a month, cutting down mature trees (Fig. 7.2) (Godai and Hashiguchi 1905). A fallen tree was cut into evenly sized strip called "Hiragi." The local people loaded piles of these strips onto "Toisan" (wooden frame backpacks) and descended the mountain with Toisan on their backs. The Satsuma Clan strictly managed the Hiragi strips and sold them for roof tiles. The local people were however able to exchange surplus strips for food such as rice distributed by the clan to enhance their well-being. Some stumps of the giant trees felled at that time are still preserved inside the forest. These buried trees are called "Domai-boku" and are a major tourist attraction at present.

During the Meiji Period (1868–1912), the control of land was stratified through the division between public and private lands. Public lands (controlled by the state bureaucracy) encompassed around 80% of the island's territory; including areas near the villages that were traditionally used as "commons" by rural societies (Inamoto 2006). Local people were prohibited from collecting fuelwood from these areas, and this caused anger in the island's communities. There were petitions for reconversion of public lands to private ownership, but the bureaucracy did not recognize many of these pleas. An example was the 1904 petition by the Kami-Yakumura and Shimo-Yakumura areas for the reinstatement of private ownership of land, but the petition was nullified by a 1920 court order recognizing state ownership of the land (Supreme Court of Japan 1920 cited in Houjou 1983, 435–466). This decision understandably did not quell public dissatisfaction, and in 1921 a new decree was passed to allow for fuelwood and charcoal collection from the coastal highlands, and provision for a subsidized price on market fuelwood and charcoal. This "Yakushima Kenpo (Yakushima Constitution)" allowed some local access to the mountain areas, and pacified the local anger to an extent. From 1922, the island witnessed a new wave of state controlled construction and timber felling in the form of the forestry railways (Fig. 7.3). In Yakushima, the forestry railways facilitated the development of a timber extraction villages such as Kosugi-dani hamlet, and a logging office directly managed by the government was established in the same area.

From the early twentieth century, the ecological value of Yakushima began to be evaluated by botanical researchers. At this time (1914), American botanist Ernest Henry Wilson visited the forest of Yakushima and introduced it as "the most interesting and remarkable forest in all Japan" (Wilson 1916, 67). During the expedition in Yaksuhima, his group found a gigantic Yaku-sugi stump, which was later named the "Wilson's stump." Its location is currently a well-known tourist attraction.

Fig. 7.2 Images of forestry during the Edo period. Source: The National Museum of Japanese History, Chiba, Japan

Fig. 7.3 A photo of a newly constructed timber railway in Yakushima during the 1920s. Source: Yakushima Forest Ecosystem Conservation Center

Because of these changes, 4300 ha of national forest on the island was designated as a "protection forest for research" in 1922, and the island highland forest was designated as a natural monument named "the Yaku-sugi Old Growth Forest" in 1924 with a recommendation of Tashiro (1923). In the first management plan of Yakushima national forests, cutting of Yaku-sugi cedars aged over 700–800 years was strictly prohibited, and in this way the old growth forest was preserved in part (Kagoshima Dairinkusho 1923).

From 1920s, the forest railway system expanded to all sections of the island. The first steam locomotive was introduced to Yakushima in 1927, helping to make Yakushima an important timber source. At this time, the felling of trees was not limited to Japanese cedar, and several broadleaf varieties were cut to supply the growing need for fuel (Inamoto 2006; Shibasaki 2015b). Continuous wars, however, cast a dark shadow on Yakushima's society gradually. Although the cutting of Yaku-sugi cedars was prohibited by FA in principle, this rule was not strictly adhered to after the early 1940s, because large amounts of timber were sequestered by the order of the Imperial Japanese Army (Tsuda 1986). There was also an increase of charcoal production in the early stage of the Pacific War (1941–1945); charcoal produced in Yakushima at that time was confidentially shipped to military factories standing outside of the island (Shibasaki 2015b). However, the US forces' continuous bombing of Japan included raids on Yakushima Island, and this led to a halt in charcoal production, and the eventual shrinkage of timber extraction (Tsuda 1986; Shibasaki 2015b).

After the war came to an end, the national forestry business in Yakushima reopened in the middle of the 1950s. In 1956, the chainsaw was introduced, leading to faster depletion of the forests. The forestry plans officially approved the cutting of the ancient Yaku-sugi cedar trees during this time (Inamoto 2006). This period witnessed an island-wide clear-cutting, and depletion of forest reserves, as the large areas of broadleaf forests around villages were also consumed. The population of the timber hamlet of Kosugi-dani reached around 550 in the 1960s. However, this acceleration eventually led to a slowdown, as forest reserves dwindled fast. Also, forest roadways were constructed, which meant that the forest railways fell out of favor quickly, and in 1970 Kosugi-dani, the last forestry hamlet, was closed (Tsuda 1986).

The closure of Kosugi-dani hamlet did not result in an immediate reduction of deforestation in the 1960s and 1970s. Forests around seashore villages were regularly felled, and timber was extracted on trucks. At this point, though, there occurred a distinct change: the first efforts to conserve the island's forests and the natural landscape began because of resistance by outside groups, and a resident group named "Yakushima wo Mamoru Kai (Group for protecting Yakushima)" (Shiba 2007). The combined pressure of such nature conservation movements became a key to the slow shift towards "forest conservation" from "forest consumption" after this period. In 1964, Yakushima was incorporated in the Kirishima National Park. In 1975, a part of the Hanayama area was designated as a Wilderness Area. In 1980, the Hanayama area (including its adjacent territories) was designated as the Yakushima Island Biosphere Reserve; furthermore, the BR was extended to the whole Yakushima Island and the property renamed as the "Yakushima and Kuchinoerabu Jima Biosphere Reserve" in March 2016 (UNESCO 2016).

In 1982, FA re-introduced the ban on cutting "Yaku-sugi" ancient Japanese cedars. In 1992, this conserved forest status was re-affirmed, and the forest ecosystem was recognized as a valuable national asset. This long history of cultural association with forests, punctuated by a period of extensive deforestation, but succeeded by nature conservation ideals, led finally to the inscription of Yakushima on the World Heritage List in 1993 (MoE et al. 2012). Subsequently, in 2009 a part of the former forest railway system was recognized as "heritage of industrial modernization" by the Ministry of Economy, Trade and Industry (Ministry of Economy, Trade and Industry 2009).

7.3.2 Changes resulting from World Heritage listing

In the 1980s, annual visitors (including local islanders who traveled to different sites) numbered around 100,000 or thereabouts, but while this changed after a high-speed hovercraft service was introduced in 1990 (Shibasaki and Nagata 1999), the WHS brand also attracted visitors, and by 2007 these had climbed to around 400,000 (Shibasaki 2015a). This change effected a new phase of socioeconomic development as Japanese style inns were built to cater for the increasing tourist demand and ecotourism became a feature of the island's tourism. Tour packages began to offer information on the natural landscape, especially centering on the ancient Jomon-sugi. It is estimated that 25% of all tourists took part in an "ecotour" while visiting the island during this period (Shibasaki 2015a). While there were only about 20 guides at the beginning when Yakushima was designated as a WHS, this number has climbed to the 160 employed at present (200 if non-registered guides are included).[6]

Beginning from the point when the island was inscribed onto the WHS list, MoE, FA, the Kagoshima Prefectural Government, and Yakushima-cho Town Office came together to build new recreation facilities and amenities such as wooden walkways, viewing decks, and public toilets. These facilities made it easier for tourists to enter the mountains and forests of Yakushima. In short, the increasing number of tourists, the demand pressure, and supply deficiencies led to a spiral of construction and burdening of the landscape (Shibasaki 2005). The increasing number of visitors, consequently, has caused issues over human waste management (Shibasaki et al. 2006). Public organizations built more toilets in congested areas and then invited larger numbers of visitors to the mountainous region (Shibasaki 2015a). A part of the donated money from the Donation System of Yakushima Mountains Conservation is meant to be used for portage of human wastes out of the listed areas; however, wastes of Jomon-sugi trekkers are sometimes left behind in the mountainous regions. Furthermore, construction and facility development has led to a complication of ownership and management systems as more and more management bodies (stakeholders) have emerged (Shibasaki et al. 2006). Although the administrative units have tried to solve this problem by creating several "coordination committees," with an eye to effective information sharing and dialog, this step has only complicated matters in effect by introducing a further loop (coordinating bodies) into the already complicated decision-making system.

It was only in the end of the 1990s that restriction of access became a focus of substantial debate. Since 2000, a shuttle bus service named "The Arakawa Trekking Bus," has been started to restrict access by private vehicles during the peak tourist season of Golden week[7] at the Arakawa Trailhead, a major visiting point for tourists who trek to see the Jomon-sugi Cedar. The Arakawa bus service was expanded to cater for tourists during summer in 2007 and later, in 2010, the bus service was opened in the period March to November every year. Although the introduction of the shuttle bus service solved the problem of congestion of the parking space at the trailhead, it did not lead to an effective

[6] Based on statistical data offered by Yakushima tourism association.

[7] The series of public holidays beginning from the end of April and ending in the first week of May.

solution of total visitor access. It was reported that peak tourist seasons in the late 2000s attract so many visitors by bus that the trail has become more congested than ever before (Shibasaki 2015a).

It was against this backdrop that the concept of restrictions on visitor access as per the Ecotourism Promotion Law (administrated by MoE) was mooted for the island in 2009. This concept involved restricting the maximum number of visitors allowed per day on the Arakawa trail to 420 (revised from the 430 as initially proposed), and the recognition of the trail area as a Specified Natural Tourism Resource (the plan was finalized in November 2009).[8] However, the idea was not well received by all stakeholders; several locals and council members complained about the financial effect (loss), and withdrew from the planned initiative. When the town office submitted the proposal to the town assembly in June 2011 it was voted down, and the proposed scheme was nullified. Meanwhile the lack of revenue for maintaining the trail remained an unsolved problem. This situation prompted new initiatives by the local municipality for collecting sufficient revenue and started to collect a token fee of 1000–2000 Yen from trekkers in March 2017.

7.4 The Problem of Decline in Cultural and Historical Significance

Multiple designations as protected areas such as the National Park in 1964 and the WHS in 1993 led to a rise in the popularity of Yakushima as a tourist destination. However, these inscriptions also brought restrictions on many customs and local stipulations. Some decisions by management authorities have not been received well by local residents. For example, a traditional custom known as "Takemairi" (which involves the cutting of the Yakushima Rhododendron, a protected species) has lasted for more than 400 years in Yakushima. In this ceremony, representatives of each community climbed up their worshiped mountains carrying sea sands and distilled spirits as offerings to mountain gods once or twice a year. Local village community representatives used to break some branches of the Yakushima Rhododendron and handed them to the villagers as a token of appreciation (Yaku-cho Town Committee for Local History 2007). But after Yakushima was included in the National Park territory, some rangers gradually treated this as an illegal activity, and subsequently this activity of breaking rhododendron branches was prohibited. It is important to note here that, although national parks are strictly protected areas, in Japan, activities that were already in progress by the time national park status was attributed are not subjected to restrictions. When an area is declared to be a special zone of a national park, all activity related to development and change of the landscape must be accompanied with a permit by MoE in principle, but a special provision called "Ki Chakushu Koui or Chakushu Koui" (ongoing activities that started before the national park status came into effect) exempts local cultural practices. In the area around Mt. Fuji,[9] for example, local organizations have rights of common land in the foothills of the mountain, and these organizations can sell licenses to collect forest by-products such as mushrooms to outsiders. However, because such examples are rare, the MoE Ranger Handbook does not enlist ongoing traditional practices in detail, and this raises the possibility that new MoE rangers may not be aware of the significance of such traditional activities on Yakushima. Furthermore, MoE officials are transferred within 2–3 years of assuming a particular post so there is a possibility that the appointment of a new ranger results in the stoppage of a traditional practice. The practice of collecting rhododendron branches should have been recognized as a cultural tradition but as this practice was carried out only once or twice a year, awareness remained low and it was interpreted as an "illegal" activity. Some of local residents started to cut branches of rhododendron again during Take-mairi to preserve their customs in recent years (Shibasaki 2016).

In addition, another value of the "remains" relating to the former forestry business in Yakushima's national forests should be utilized as tourist attractions. As explained before, the intensive clear-cutting system using forest railways ended, and forestry villages in national forests were closed in 1970. But many forestry facilities such as dwelling sites are literally "sleeping" in Yakushima's forests (Shibasaki 2015b) (Fig. 7.4).

Many of these remains, however, face the threat of damage, and some have been lost by forestry thinning operations already. The reason behind this is that MoE or FA make their own plans for park and forest management, and manage "resources" accordingly. However, the "remains" of old forestry or similar infrastructure items that have no ongoing use value are not listed as a "resource." Thus, although visitor centers and walkways are listed as "heritage properties," the remains of the past forestry industry are not included in that list. In effect, such

[8] When an area is designated as a Specified Natural Tourism Resource, the Mayor of a Municipality may make such restriction as to require prior approval of said Mayor for any entry in the area where said Specified Natural Tourism Resources are located, pursuant to the provisions of Ordinance of the competent Ministry (Article 10(1), Ecotourism Promotion Act).

[9] Mt. Fuji, designated as a "cultural" WHS in 2013, has natural resources local people have utilized as well as prominent landscapes, and was one of the earliest Japanese national parks designated in 1936. Local NGOs and prefectures promoted Mt. Fuji to be designated as a "natural" WHS in the 1990s; however, this plan fell through partially due to overuse issues, and the promotion of the mountain as a "cultural" WHS was started in the 2000s as an alternative (Sano 2008).

Fig. 7.4 Remains of forest railways and dwelling sites. Source: Photo by author

historical artifacts are "nonexistent" on official maps. These forestry remains and associated people's memories should be evaluated more highly to know how FA and ex-workers utilized (overexploited) the diverse Yakushima forests, which may in turn deepen understanding of the environmental history of Yakushima (Shibasaki 2015b). If these forestry remains and people's memories could be conserved, they would be precious study materials for forestry or environmental education.

7.5 "Simplification of Values" as an Issue

It is necessary to point out that the issue of "simplification" of values is a matter of concern for Yakushima Island. As Scott (1998) showed, policymakers tend to pay attention to the single value of forests to accomplish targets related to their main management and socioeconomic purposes. Thus, between the 1920s and the early 1980s, the "value" of Yakushima was mainly focused on the forestry made up from Yaku-sugi timber and charcoal. The introduction of chainsaws devastated the landscape of mountains though large-scale clear-cutting leading to local protest against FA's ways of forest management. From the early 1980s, cutting of Yaku-sugi was banned, and forest management policies dramatically changed to being dominated by conservation. Instead of forestry values, the values of ecology and landscape became dominant, and this direction accelerated after the WHS. This condition is depicted in Fig. 7.5, where different "values" are expressed with different colors, and the total resource value of a given area is measured in terms of the area of the colored patches.

Once the area is given a special "protection" or "heritage" status, only one part of the heritage receives disproportionate attention leading to the strengthening of efforts to preserve that aspect, and increased media focus on that one type of "value." In extreme cases, over time, other types of "values" could be extinguished from the landscape. In addition, as time goes on, the one value that was highlighted during the enlisting process can become more and more independent as it loses connection with related sociocultural aspects and eventually loses attraction, driving down its associated tourism capital in the process. Also, excess commodification of the main value may produce skewed value distribution (see the attached diagram for illustration of this point).

Thus, if the graphic in Fig. 7.5 is used to explain the situation in Yakushima, the main value that is highlighted during the registration process will be the landscape and unique ecosystems. The "other values" would be forestry practices and local cultural traditions which are currently obscured. WHS registration led to the peaking of tourist interest on the Jomon-sugi cedar, and this became the main tourist capital of the island. Public organizations promoted more visitors to come to the mountainous region by building boardwalks and toilets; however, these public services caused issues of congestion and human waste. Also, FA banned Yaku-shika deer hunting in national forests in principle, and this could be one of the reasons for the population explosion of this species.

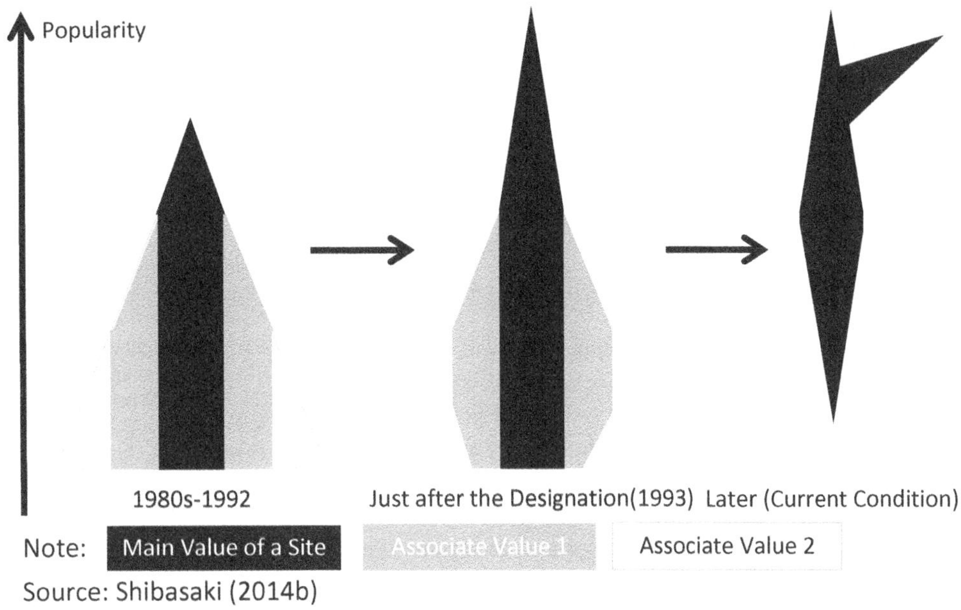

Fig. 7.5 Changes in the value of protected areas after enlisting or registration as a World Heritage site

Furthermore, it must be noted that this form of simplification is not only fostered by public organizations. In the case of Yakushima Island, ecotourism and the media industries also strengthened the "simplification" (Shibasaki 2014b). For instance, the "Wilson's stump" was introduced to visitors with the story of Wilson's expedition and the framework of local worship for the stump. After designation, however, a viewing place for the stump was found where the shape of the cavity is "heart-shaped." This heart-shape cavity came to be introduced as a lucky symbol by ecotourism guides and TV broadcasters, and some trekkers visit Wilson's stump to see the heart-shaped cavity without knowing the history of the site at all. Commodification of the Wilson's stump has produced added value, but this has no relation with Wilson's achievement at all.

Heritage in a Nutshell: Yakushima

Type of Heritage: Composite: Ecological with Geological and Geomorphological characteristics.

Yakushima became UNESCO World Heritage (Natural Area) in 1993, as one of the first World Heritage (Natural) properties from Japan. The island, located nearly 135 km from Kagoshima Port in Kyushu, is noted for its rich flora (containing nearly 1900 species and subspecies, a montane temperate rainforest, and very large specimens of old growth Japanese cedar); the diversity of marine fauna off the coast; and vertical distribution of its ecosystems. The island is also the location of a UNESCO Biosphere Reserve.

While the perception of Yakushima as a pristine wilderness landscape is strong, the forests have a very long association with human activity. Oku-dake mountains were considered sacred landscapes in premodern times and were rarely visited before the seventeenth century. However a major wave of forest cutting ensued afterwards, and Yakushima became a major forestry resource, a phase that continued until the early 1980s. This social history of Yakushima's forests is currently not highlighted adequately; and understanding of cultural and historical significance can deepen the understanding of environmental history of this heritage landscape.

Tourism: Yakushima Island and the sea around it are locations for a number of nature-based tourism or ecotourism activities. Forest trails are highly popular; the trail to the Jomon-Sugi Cedar is a major highlight. Hiking mountains, kayaking, and diving in the sea and river kayaking are also popular activities. A number of lodging facilities are available along the coast. However, tourism adds pressure to the sensitive forest landscape and should be conducted in a responsible manner.

Useful websites:

Yakushima Tourism Association http://yakukan.jp

MoE Yakushima National Park https://www.env.go.jp/en/nature/nps/park/yakushima/index.html

This ongoing simplification and fragmentation of value is affecting the economics of the island communities as well, as it is associated with the decline of visitor numbers to Yakushima Island after 2010 (Shibasaki 2015a). After world heritage inscription, the island landscape became dominated by tourism, but this boost was short-lived (Shibasaki 2015a). As construction and other types of business opportunities declined, tourism became even more prized for its economic role, but tourist numbers gradually declined. The current situation is that of stasis where new development pathways are unclear.

7.6 Conclusion

This chapter discussed the landscape history of Yakushima, focusing on how Yakushima's forests have been modified by human activities. The island rose from granitic exclusion in geological time and orogenic activity and volcanism have shaped the landscape over time. Yaksuhima is recognized as a significant natural heritage due to its vertical ecosystem distribution, an important resting and breeding point for migratory species (several of which are endangered) and for landscapes of visual beauty (forests and mountains). However, the forest has been heavily used over the centuries; external bodies dominated the forestry practices on Yakushima for a long time. Currently even as the island is protected under the world Heritage convention and national park stipulations, tourism-related drivers have pressurized parts of the landscape A skewed value distribution favoring specific landmarks adds to this pressure and there is a possibility that a disconnect between geology, ecology, and local culture could occur. In this regard, the issues of "simplification" of values in Yakushima Island should be deepened. Of course, the issues of simplification is not peculiar to the Yakushima society, and some case studies can be seen such as competing values of Great Smoky Mountains National Park (Young 2006) and divergence of spatial recognition between the Thai state and local people (Roth 2008). Although there is no panacea to solve this issue, values of Yakushima and a grand design for whole Yakushima should be fundamentally re-examined by broad stakeholders including researchers as well as silent majority of local people (Shibasaki 2015c). In particular, community-based values (getting back a yardstick of Yakushima's people) may rebalance conservation policies of Yakushima. The concerned authorities also need to pay close attention to the connectivity of natural systems (such as geology, geomorphology, and ecology and the integrity of processes operating in those domains) and local cultural specificities to holistically manage this truly remarkable heritage.

Acknowledgements This research was supported by the JSPS KAKENHI Grant Number (16H04940). The author thanks the editors of this volume for their assistance and suggestions on the manuscript preparation.

References

Godai H, Hashiguchi K (1905) Sangoku Meisho Zue 50 (Pictures of Scenic Beauty in the Three Countries 50), 1–46. http://dl.ndl.go.jp/info:ndljp/pid/992147. Accessed 26 Feb 2017

Hayashi F (1951) Ukigumo (Floating clouds). Kadokawabunko, Tokyo

Houjou H (1983) Daishinin Saikousaibansho Iriai Hanketsusyu Dai 5 Kan (Administrative court collections of Iriai adjudication Series No. 5). Ochanomizu Shobo, Tokyo

Inamoto T (2006) Yakushima Kokuyurin no Segyoushi (Management History of Yakushima National Forests). In: Osawa M, Yamagiwa J, Tagawa H (eds) Sekaiisan Yakushima: Anettai no Shizen to Seitaikei (World Heritage Yakushima: nature and ecology of sub-tropical areas). Asakura Shoten, Tokyo

IUCN (1993) World Heritage Nomination–IUCN Summary 662: Yakushima (Yaku-Island) (Japan). http://whc.unesco.org/document/153977. Accessed 26 Feb 2017

Kagoshima Dairinkusho (1923) Yaksuhima Kita Higashi Minami Jigyouku Segyouan Setsumeisho (Management plan for Yakushima National Forests of Northern, Eastern, and Southern project segments). Kagoshima Dairinkusho, Kagoshima

Ministry of Economy, Trade and Industry (2009) Kindaika Sangyo Isangun Zoku 33 (33 Heritage constellations of industrial modernization Vol. 2) http://www.meti.go.jp/policy/local_economy/nipponsaikoh/pdf/isangun_zoku.pdf. Accessed 26 Feb 2017

MoE (Ministry of the Environment) (2007) Red list category. http://www.env.go.jp/press/file_view.php?serial=9953&hou_id=8648. Accessed 26 Feb 2017

MoE (Ministry of the Environment), the Forestry Agency, the Agency for Cultural Affairs, Kagoshima Prefecture, and Yakushima-cho Town (2012) Yakushima Sekaiisan Tiiki Kanri Keikaku (Yakushima World Heritage area management plan). https://www.env.go.jp/park/yakushima/ywhcc/wh/wh_4_01.pdf. Accessed 26 Feb 2017

Okano T, Matsuda H (2013) Biocultural diversity of Yakushima Island: Mountains, beaches, and sea. J Mar Island Cult 2(2):69–77

Osawa M (2006) Yakushima no Shokuseibunpu (Natural vegetation in Yakushima Island). In: Osawa M, Yamagiwa J, Tagawa H (eds) Sekaiisan Yakushima: Anettai no Shizen to Seitaikei (World Heritage Yakushima: nature and ecology of sub-tropical areas). Asakura Shoten, Tokyo

Roth R (2008) "Fixing" the forest: the spatiality of conservation conflict in Thailand. Ann Assoc Am Geogr 98(2):373–391

Saito M, Ogasawara M, Nagamori H, Geshi N, Komazawa M (2007) Geological map of Japan 1:200,000, "Yakushima". Geological Survey of Japan, AIST. https://www.gsj.jp/data/200KGM/PDF/GSJ_MAP_G200_NH5209_2007_D.pdf. Accessed 26 Feb 2017

Saito M, Geshi N, Ogasawara M, Nagamori H, Komazawa M (2008) 'Kankochishitsugaku' karamita Sekaiisan Yakushima: 1:20000 Chishitsu Zuduku 'Yakushima' no Kanko. (World Heritage Yakushima in a Viewpoint of Sightseeing Geology: New publication of Geological Map of Japan, 1:200,000 Yakushima). Chishitsu News 647: 2–60. https://www.gsj.jp/data/chishitsunews/08_07_09.pdf. Accessed 26 Feb 2017

Sano M (2008) Fujisan no Sekaiisan Touroku heno Torikumi niokeru Shizenisan toshiteno Kachihyouka no Igi (Significance of evaluation of Mt. Fuji-san as World 'natural' heritage in the movement for designation as the World heritage). Tikyu Kankyo 13(1):51–60

Sato T, Nagashima H (1979) Geology of the Seinanbu district in Yakushima. Quadrangle Series scale 1:50,000. Geological Survey of Japan. https://www.gsj.jp/data/50KGM/PDF/GSJ_MAP_G050_16009_1979_D.pdf. Accessed 26 Feb 2017

Scott JC (1998) Seeing like a state: how certain schemes to improve the human condition have failed. Yale University Press, London

Shiba T (2007) Ano Junen wo Kataru: Yaku-sugi Genseirin no Hogo wo Megutte (Talking about the 10 years: for protection of Yaku-sugi Virgin Forests). Goyoshobo, Tokyo

Shibasaki S (2004) Yakushima no Kankou Boom wo Kangaeru (Comments to a tourism boom to Yakushima Island). Sanrin 1445:29–35

Shibasaki S (2005) Kiro ni Tatta Yakushima no Kanko (Yakushima's tourism standing at a crossroad). Shinrinbunka Kyokai, Tokyo

Shibasaki S (2014a) Sekaisizenisan no Saishigenka nimuketa (Action research on the sustainable resourcing of world heritage). A report on the research achievements of JSPS Kaken Grant (23310035) https://Kaken.Nii.Ac.Jp/Ja/File/Kakenhi-Project-23310035/23310035seika.Pdf. Accessed 26 Feb 2017

Shibasaki S (2014b) Hogochiiki no Touroku Shitei ga Tikishakai ni Oyobosu Eikyou: Yakushima wo Jirei toshite (Impacts of designation as protected areas to local communities: a case of Yakushima). Sonraku Shakai Kenkyu 21(1):45–48

Shibasaki S (2015a) Yakushima niokeru Ecotourism-gyo no Keizaibunseki (Economic analysis of the ecotourism industries in Yakushima Island). Bull Natl Mus Jpn Hist 193:49–73

Shibasaki S (2015b) Yakushima no Mori ni Nemuru Hitobito no Kioku (People's memories sleeping under forests of Yakushima Island). National Museum of Japanese History Ethnological Research Film

Shibasaki S (2015c) Nyuzanryo wo Toreba Nyuzankisei wo Okonaeba Yakushima no Sangakuriyou Mondai ha Kaiketsu surunoka? (Are overuse issues in Yakushima Island solved by implementing of a quota as well as fee system?). Kankou Bunka 226: 9–25 https://www.jtb.or.jp/wp-content/uploads/2015/07/bunka226-06.pdf. Accessed 26 Feb 2017

Shibasaki S (2016) Fukkatsu shita Takemairi (Reviving Takemairi Custom). National Museum of Japanese History Ethnological Research Film

Shibasaki S, Nagata S (1999) Estimation of the number of tourist to Yakushima Island. Jpn For Econ Soc 45(3):33–38

Shibasaki S, Hirata K, Yokota Y, Nagata S (2006) Sekaishizenisan Touroku ga Tiikishigen Kanritaikei ni Oyobosu Eikyou (Influences of registrations of the World Natural Heritage on the system of natural resource management: a case study of the mountainous region in Yakushima Island). Shinrin Keizai 59(8):1–16

Supreme Court of Japan (1920) Kokuyusanrin Sagemodoshi Shinseikyakkashobun Torishimari no Uttae (Judges' decision against the claim of Kami-yaku-mura and Shimo-yaku-mura villages for taking monuments of Yakushima Island, Kagoshima Prefecture)

Tashiro Z (1923) Kagoshimaken Yakushima no Tennenkinenbutsu Chousa Houkokusho (A research report on Natural monuments of Yakushima Island, Kagoshima Prefecture). Reprinted in 1995 by Seimei no Shima, Kagoshima

Tsuda K (1986) Yaku-sugi ga Kieta Tani (Valleys where Yaku-sugi Cedars were cut down). Asahi Shinbunsha, Tokyo

UNESCO (2016) Final Report, International Coordinating Council of the Man and the Biosphere (MAB) Programme 28th session, Lima, Peru,18–19 March 2016. http://www.unesco.org/new/fileadmin/MULTIMEDIA/HQ/SC/pdf/SC-16-CONF-228-12_final_en_v2.pdf. Accessed 26 Feb 2017

Wilson EH (1916) The conifers and Taxads of Japan. Cambridge University Press, Cambridge

Yaku-cho Town Committee for Local History (2007) Yaku-cho Kyoudoshi: Shizen Rekisi, Minzoku, Vol. 4 (Local history of Yaku-cho Town Vol. 4: Nature, history, and folklore). Yaku-cho Town, Kagoshima

Yakushima Environmental Culture Foundation (2000) Diagram Yakushima. Yakushima Environmental Culture Foundation, Kagoshima

Yakushima World Heritage Conservation Center (nd) Umigame no Sato: Nagata-hama nitsuite. (A village for sea turtles: About Nagata-hama). https://www.env.go.jp/park/yakushima/ywhcc/np/nagata.htm. Accessed 26 Feb 2017

Yakushima-cho Town (2016) Chohou Yakushima (Yakushima Public Information) 106.http://www.town.yakushima.kagoshima.jp/t_yakushima/wp-content/uploads/2016/07/17c44876957d8bd2e40fb794ef23ab77.pdf. Accessed 26 Feb 2017

Young T (2006) False, cheap and degraded: when history, economy and environment collided at Cades Cove, Great Smoky Mountains National Park. J Hist Geogr 32:169–189

Part II

UNESCO Global Geoparks

The Origin and Development of Geoparks in Japan: Reflections from a Personal Perspective

8

Mahito Watanabe

Abstract

This chapter provides a historical overview of geopark development in Japan, especially focusing on the initial stage. The author was a key stakeholder in these efforts, and the chapter is based on his personal experience. In the initial stage, excepting a few geoscientists, the geopark concept was virtually unknown. The International Year of Planet Earth (IYPE) was a major catalyst for introducing geopark activities into Japan. The first key stakeholders were academic societies and scientists, and it was realized at that stage that local administrative participation was vital for the growth of this concept in Japan. The Japan Geopark Committee (JGC) was established to evaluate the scientific merit of applications, and the first designated geoparks voluntarily set up the Japanese Geoparks Network (JGN). The first global geoparks from Japan were recognized in 2009, and this helped generate media attention and popular interest. Today, Japan has 35 national geoparks and 8 UNESCO global geoparks. The activities have become much more diverse compared to the initial years, the stakeholder base have become broader, and educational programs in geoparks have gained praise from evaluators; however, several issues such as conservation of geological heritage and long-term plan for sustainable development needs more attention.

Keywords

Geological heritage • Sustainable development • Geopark movement • Earth science for society • UNESCO • IYPE • JGC • JGN

8.1 Introduction

Geoparks have become increasingly popular in Japan over the years. The Geopark concept is fundamentally related to conservation of geoheritage and sustainable development (Zouros 2004; McKeever and Zouros 2005). The geopark movement began in Europe in the 1990s, and the Global Geoparks Network (GGN) was established as an autonomous networking body for international geoparks in 2004; with European and Chinese geoparks as first members, with support from UNESCO (Eder and Patzak 2004). The geopark movement was supported by UNESCO from an early stage, but it did not become an official UNESCO program until 2015, when all existing global geoparks were incorporated into the newly created International Geoscience and Geoparks Programme (IGGP). Throughout their history, global geoparks have contributed to the promotion of earth sciences, the conservation of internationally significant geological heritage; and they have created a bridge between earth science and society by providing a platform for interpreting the planet's heritage. Geoparks have also created new tourism resources and have utilized earth science for raising awareness and preparedness for natural

M. Watanabe (✉)
Geological Survey of Japan, National Institute of Advanced Industrial Science and Technology (AIST), Tsukuba, Japan
e-mail: mht.watanabe@aist.go.jp

A. Chakraborty et al. (eds.), *Natural Heritage of Japan*, Geoheritage, Geoparks and Geotourism,
DOI 10.1007/978-3-319-61896-8_8

hazards. The first global geoparks from Japan emerged in 2009, and since then several geoparks have successfully attained the UNESCO Global Geopark (UGG) status. At present, 8 UGGs exist in Japan. In addition, 35 national geoparks currently exist in Japan (Fig. 8.1). In this chapter, I reflect upon the initial phase of geopark promotion in Japan and provide an overview of the maturity of the geopark initiatives over the years. As I am involved with the diffusion of the geopark concept in this country since the very first stage, the narrative that follows is primarily based upon my own direct experience as a scientist, a geopark specialist, and through my duties as an evaluator of geoparks and their heritage.

8.2 Geoparks: Concept and Structure

UNESCO Global Geoparks are, as defined in the Statutes of the International Geoscience and Geoparks Programme, "the mechanism of international cooperation by which areas of geological heritage of international value, through a bottom-up approach to conserving that heritage, support each other to engage with local communities to promote awareness of that heritage and adopt a sustainable approach to the development of the area" (UNESCO Global Geopark Guideline 2015).

In other words, geoparks are areas where *natural heritage* is based on geological or geomorphological characteristics; however, merely possessing important geological landmarks is not enough for a region to qualify as a geopark. To be successfully recognized as a geopark, the concerned area should be a model for sustainable development through the promotion of its geological/geomorphological uniqueness (The Shimabara Declaration, adopted at 5th International UNESCO Conference on Geoparks 2012). The most fundamental criteria for becoming a successful geopark is: conservation of the natural and cultural heritage of the concerned area, education about the value of that heritage, promotion of earth science-related knowledge, and efforts to create new tourism opportunities by communicating the value of geological heritage to visitors (UNESCO IGGP 2015).

The most important aspect of a geopark is the *connectivity* between the geological history of Planet Earth as a whole

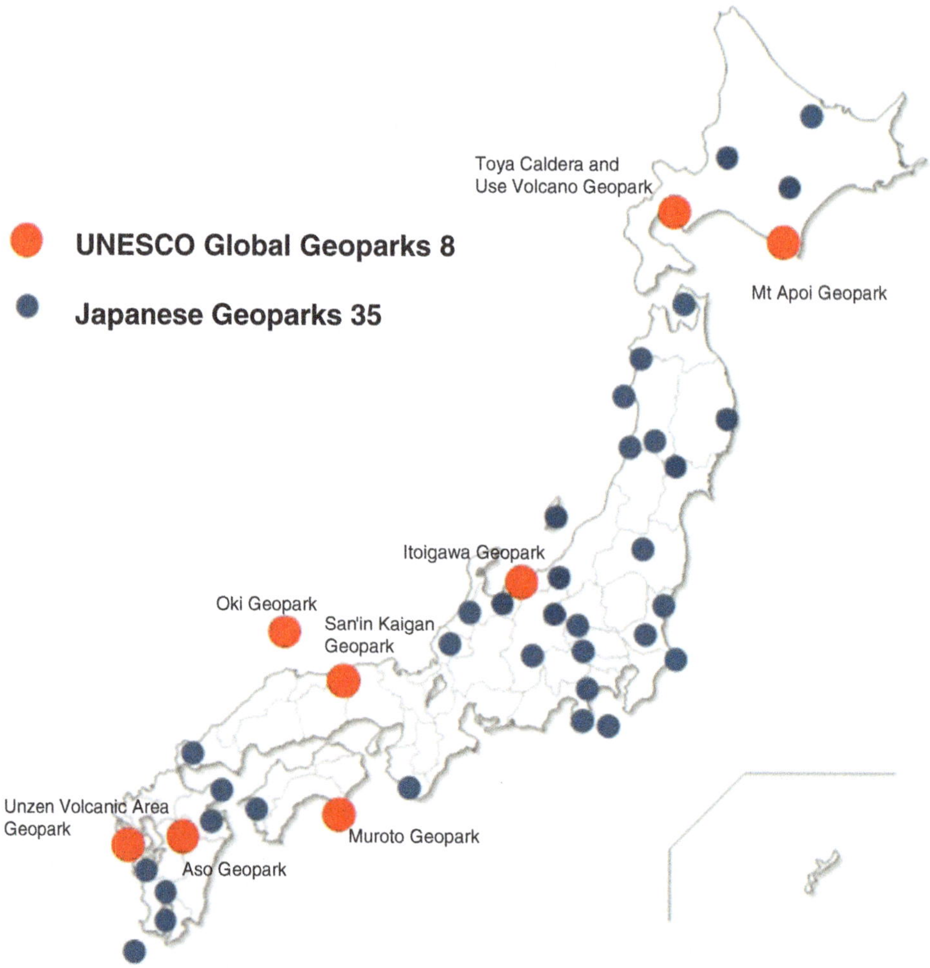

Fig. 8.1 Location of geoparks in Japan (courtesy Japanese Geoparks Network)

and the geological or geomorphological attributes of the given area's natural heritage. The *geo-story* is based upon a narrative that explores this connectivity through different aspects such as geological landmarks, events, and social or cultural attributes. Although geoparks are based on the concept of geological or geomorphological uniqueness and its protection, the concept of geopark is not only limited to geology and geomorphology and includes the given area's ecosystems, culture, and traditions. Therefore, geoparks should make efforts to promote these aspects along with promoting geological uniqueness. Tourism is of fundamental importance to a geopark's economy, geoparks should promote responsible and sustainable tourism that aids the protection of the natural environment. The IGGP statues state that it is expected that geoparks simultaneously contribute to geo-conservation and economic development (UNESCO IGGP 2015). Another important aspect of a geopark is that it must be a "bottom-up" initiative involving citizens and local administration (Ibid). The UNESCO guideline for global geoparks explain in detail about the various evaluation criteria and about how geoparks can contribute to conservation of the valuable geological or natural heritage, sustainable economic development, promotion of earth science awareness among both local residents and visitors, awareness of environmental problems, and resilience in the face of natural hazards (Ibid).

Geoparks became an officially recognized UNESCO initiative in 2015. However, the international network of geoparks began its journey back in 2004, in the form of the Global Geoparks Network (GGN). Prior to this, the concept was born in Europe in the 1990s, and in 1997 there was an early effort to nominate geoparks as an official UNESCO program. However in 2001, it was decided that geoparks would not gain the official UNESCO status for the time being, and UNESCO would support the program on an independent basis. It was after this development that global geoparks began to be certified under the GGN (an autonomous body). In 2015, it was recognized that the GGN's activities and the activities of each of the global geoparks have contributed significantly to the conservation of geological heritage and bottom-up local development; and the initiative was awarded a formal UNESCO program status.

8.3 A Brief Outline of Geopark Development in Japan

8.3.1 The Geological Setting

The Japanese Islands are located along the western rim of the Pacific Ocean, which is one of most active mobile belts on the earth. In the past, these islands were affected various kinds of geo-tectonic movements caused by subduction or collision of plates (e.g., Taira et al. 2016). Such geological background has helped in the wide acceptance of the geopark concept and the rapid growth of JGN as described above. The Japanese Islands and their geology have been formed through the long history of five hundred million years. The Japanese Islands were once parts of small continental blocks that formed the Eurasian landmass through collision and amalgamation about 250 Ma. Japanese Islands were separated from the continental landmass by the formation of the Sea of Japan about 20–15 Ma. During this process, the Japanese Islands have been continuously affected by subduction, tectonic uplift, and tectonic erosion. Due to the long and complex geohistory, the Japanese Islands have a rich geological diversity which is represented by various types of rocks and geological layers that span 500 million years. In addition to the geological diversity, active geological processes such as volcanic eruption, earthquake, and landslides caused by the plate motion continually enrich the diversity of landscape. In some cases, geological landforms or landscapes are regarded as sacred places and some of those landforms are utilized as tourist attractions. However, it is also observed that till date the full potential of geoheritage has not been fully recognized and utilized as educational tool or tourist attraction in Japan. Through the geopark concept, local people could notice the wonder and value of geological nature and geoheritage, and the concept obtained enthusiastic support from local administrative units and people. In this sense, the geopark concept can be claimed as an epoch-making one to foster the recognition of the rich geodiversity of Japan.

8.3.2 2004–2006: First Introduction of the Geopark Concept

A chronological history of the early phase of geopark activities in Japan is provided in Table 8.1. For a more detailed explanation, refer to Watanabe (2014).[1] The initial exposure to the geopark idea occurred through international conferences on geology, and initially research organizations that specialize in geological matters such as the Geological Society of Japan, an NPO with the acronym GUPI[2] and the Geological Survey of Japan (GSJ) (one of research institutes of the National Institute of Advanced Industrial Science and Technology (AIST)). A detailed description of the geopark activities between 2004 and 2007 can be found in Iwamatsu (2008) (Fig. 8.1).[3]

[1] In Japanese.

[2] This NPO specializes on promoting wider use of geology-related data.

[3] In Japanese.

Table 8.1 History of the recognition of the Geopark concept in Japan

2000	The European Geoparks Network (EGN) is established through support of the UNESCO and EU
2002	The International Geoscience Programme (IGCP) of UNESCO decides to promote geopark activity
2004, March	Chair of the Japan National Committee for IGCP, Shigeki Hada conducts lectures at San'in Kaigan area
2004, April	Geological Information Utilization and Promotion Initiative (GUPI) is founded
2004, July	GGN is established through the support of UNESCO
2004, August	GUPI devotes a section of its homepage for geoparks
2004, September	A small gathering takes place at the Geological Society of Japan regarding IYPE
2005, October	Establishment of Geopark promotion Committee in the Geological Society of Japan
2006, September	Three presentations on aspiring Japanese geoparks at the second UNESCO Geoparks Conference in Northern Ireland
2007, January	Initiation of events regarding the IYPE (2008), events continued into 2009
2007, May	The first Geopark session is organized at the annual Japan Geoscience Union (JpGU)
2007, June	Asahi Shimbun daily features a report on the JpGU geopark session, the report is widely read, leading to an increase of interest among stakeholders
2007, July	AIST publishes a special issue on geoparks
2007, October	Coordination Committee for Japanese Geoparks is organized at GUPI
2007, December	The inaugural meeting of the Japanese Geoparks Coordination Committee is held
2008, January	Coordination meeting between relevant ministerial representatives held, JGC becomes the evaluation representative for territories from Japan (Secretariat located at AIST)
2008, May	Inaugural meeting of the JGC
2008, October	JGC nominates Toya-Usu, Itogawa, and Shimabara (Unzen) as aspiring GGN members
2008, November	Toya-Usu, Itoigawa, Unzen, Muroto, San'in Kaigan, Minami Alps awarded Japanese national geopark status
2009, February	Commemorative event for certifying Japanese national geoparks, pledge to set up JGN
2009, May	Inaugural meeting of the JGN
2009, August	Toya-Usu, Itoigawa, and Unzen become the first Global Geoparks from Japan

8.3.3 2005–2007: Early Phase (Preparation of Establishment of the Japan Geopark Committee and Japanese Geoparks Network)

In 2005, a promotion committee for establishing geoparks was set up under the Geological Society of Japan. The headquarters were located inside the AIST, and I assumed responsibility for the official duties of this organization. In December 2005, the UN decided that the International Year of Planet Earth (IYPE) would be observed in 2008 and that the UN would be involved in the promotion of earth sciences during the period of 2007–2009. At the foundational geopark promotion committee in Japan, it was decided that the IYPE was a good opportunity to promote the geopark concept in the country and possibly prepare some candidacies for GGN membership. At first, those who were involved in promotion of geopark activity in Japan thought that geoparks could be promoted as a program of a relevant national level ministry, and other committee members and I spoke at length with relevant ministerial representatives in 2006. But as discussed earlier, at that time, the geopark initiative was not an official UNESCO program; and this proved to be a major stumbling block for our negotiations.

In 2006, geopark symposiums were organized by GUPI in Tokyo with the aim of familiarizing the geopark concept. This was the first time when symposiums on geopark-related themes were held in Japan. By this time, events related to the IYPE also featured some geopark-related contents; this aided the gradual diffusion of the concept among the geological community. In September 2006, four participants from Japan (including me) took part in the Second UNESCO Geoparks Conference in Belfast, Northern Ireland. During this conference, presentations on aspiring GGN members from Japan were conducted, and we had the opportunity to exchange opinions with our international colleagues who were involved in geopark promotion.

Later on, those four participants became the key members of the nascent geopark committee in Japan, and a lot of discussions took place on how to promote geoparks in the country. From these discussions, it became clear that a viable way to promote geoparks in Japan could be through involving local administrative organs and resident researchers; and eventually the base of activities could be broadened to include residents and national level stakeholders. It was realized that to manage successful geoparks and to carry out activities such as conservation, research, education, and promotion at the local level, it was of utmost importance that local level stakeholders were convinced of the value of geoparks because the activities would have to be supported by local level administrative funds. We also had the realization that it would not be feasible to promote geoparks in Japan through the support of private partners or scientific organizations alone as awareness level regarding the importance of earth sciences remains low in the Japanese society. Thus, Japanese geoparks emerged as initiatives led by local administrative bodies and this situation persists today. However, at that point, the two major emphases were on the promotion of earth science-related knowledge and the economic development of the concerned regions, and the need for preserving the geological heritage and the protection of the natural environment were not adequately addressed. Detailed information about the contents and aspirations of geopark activities of this stage can be found in Tsukuda (2007), Iwamatsu (2007), and Watanabe (2007a, b), which are accessible through the AIST homepage (https://www.gsj.jp/publications/pub/chishitsunews/news-contents.html).

In May 2007, a geopark session was organized for the first time in the annual Japan Geoscience Union (JpGU), the largest annual gathering of earth scientists in the country. The geopark promotion committee played a key role in organizing this session. In this session, presentations were made on geopark activities by researchers, museum curators who were responsible for showcasing a part of their geological heritage and administrative officials. In addition, some presentations brought out the independent views of some of these stakeholders. The session's proceedings were reported in the Asahi Shimbun, one of the most widely circulated newspapers in the country. GSJ published a special edition of its public relations journal to review current state of geoparks and it was distributed among the audience; it raised the attention of public and private sector stakeholders who were in a position to liaise with the geopark activities through partnership with the AIST. It would not be an exaggeration to note that this publicity was vital; it was a major stepping stone for us as it made our activities known on a mass scale. This in turn paved the way for increased awareness and interest among the public, private, and individual stakeholders. The newspaper coverage of geopark activities continued till 2008, and looking back, we realize that this phase was a major turning point in the diffusion of the geopark concept. However, while most of the reports publicized geoparks as the geological version of UNESCO's World Heritage and a tool for local development relatively little attention was paid to the need to conserve the geological/geomorphological heritage and the need to make future generations aware of this value through appropriate education. Reflecting on this, I would also mention that at that time key members like me did not have sufficient understanding about the totality of the geopark concept and we lacked the appropriate tools to communicate the essence of geoparks to the broader audience.

Although there were several limitations in these early efforts, over ten territories and stakeholders expressed their willingness to establish geoparks by the summer of 2007. At this stage, the candidates had vastly different backgrounds. Some territories had very little understanding of the concept, some territories had some success as far as conservation and education were concerned but fared rather poorly when it came to geotourism, and some areas were inhibited by the lack of geoscience education programs. For example, Itoigawa City had taken conservation measure to some of the important geosites in the city such as the Jade Gorge and one sectional outcrop of a fault, but those sites were visited by very limited visitors. Another case is the San'in Kaigan area where beautiful coastal landscape was an important touristic attraction but the scientific significance of the rocks found in the coastal area were largely unknown to people. Eventually 11 territories came together to establish a "Coordination Committee for Japanese Geoparks" (preparatory commission for Japanese Geoparks), this was the prototype of the Japanese Geoparks Network (JGN) of today. Naturally, there were many challenges before this young coordination committee. Some members were apparently solely motivated by the global brand of geopark and did not understand the finer details. To make matters complex, geoparks were still growing in the international stage at that time; as a result, it was a situation akin to adjusting one's objectives to a shifting target. However, one major achievement of this coordination committee was that it paved the way for information exchange regarding geopark activities in a wide base of stakeholders.

8.3.4 2008-Present: Developing Phase (Establishment and Growth of JGC and JGN)

Several member territories of the coordination committee were interested in GGN membership. The GGN rules stipulated that aspiring members should be able to provide official letters of support from concerned ministries. However, as mentioned before, there was no ministry in Japan that was directly involved with geopark promotion at that stage; this was the biggest hurdle for aspiring geoparks from Japan. This problem was eventually solved with the establishment of the Japan Geopark Committee (JGC) in 2008: the committee was an independent body of scientific experts; and relevant national ministry representatives would liaise with it as "observers." The Secretariat was located at the AIST. JGC members were chosen from five academic societies: Geological Society of Japan, the Association of Japanese Geographers, the Volcanological Society of Japan, the Seismological Society of Japan, and the Japan Association for Quaternary Research. The first JGC session was held in 2008, and operational guidelines for evaluating aspiring geoparks were finalized. Accordingly, the JGC conducted its first evaluation in the summer of 2008. The first GGN aspiring members were nominated in September 2008 and in November 2008 the first Japanese national geoparks came into existence.

At that time, a maximum of three slots were allowed per country for GGN membership. The initial applicants were Toya-Usu, Itoigawa, San'in Kaigan, Muroto, and Shimabara (Unzen). Eventually, based on their activities, Toya-Usu, Itoigawa, and Shimabara (Unzen) were nominated for the GGN membership slots. Toya-Usu was recognized for its exemplary effort to preserve the memories of the 2000 eruption of Mount Usu, by keeping damaged structures in place and managing a museum that provided useful knowledge about volcanic hazards. In Itoigawa, there were early efforts to preserve important rock outcrops from 1991[4] and promote the area's geological heritage through geo-trails and explanation panels. In addition, the Fossa Magna Museum provides an excellent opportunity to learn about the area's geology and volcanism-related aspects. Shimabara was recognized for its efforts to communicate volcanic hazards and the involvement of earth scientists in science communication.

In addition to these three aspiring GGN members, four other territories, Muroto, San'in Kaigan, Minami Alps (South Japan Alps), and Mt. Apoi were awarded the status of national geoparks of Japan. The seven national geoparks came together to form a network—the Japanese Geoparks Network (JGN)—in February 2009. The headquarters of the JGN, managed by the Itoigawa Geopark, is currently located in Tokyo. Over the years, the JGN has become not only a platform for coordination between geopark stakeholders, but it has become an association that promotes geopark-related activities and various surveys to monitor the condition of geoparks in Japan. The JGN holds annual conventions featuring all Japanese geoparks and organizes workshops, seminars, and lectures by international geopark experts and managers. Currently, the JGN also manages several "working groups," which are independent forums for debate, information dissemination, and PR activities involving geoparks, geological heritage, education, and tourism-related aspects. Some of these working groups feature independent academic experts and researchers. Although these groups do not have a formal mandate, there are ongoing efforts to improve geopark activities by liaising with experts and local stakeholders.

In the summer of 2009, the three aspiring GGN territories underwent field evaluation. While evaluators favorably judged education-related initiatives, several problems were pointed out. Two major problems were: overuse of scientific terminology in pamphlets, explanation panels and tours; and the lack of independence of managing bodies. While all three territories had a broad base of local stakeholders, decision-making mechanism, strength of the promotion bureau's main office, the condition of long-term plans became points for criticism. In the end, all three territories successfully gained GGN membership in August 2009, but several problems such as the lack of tours for the general public, lack of qualified guides (apart from resident researchers), and lack of concrete planning for sustainable economic development, remained. With their successful bid to become GGN members, these three territories gained increased media attention which in turn raised the level of interest about geoparks among many more territories.

The following year, JGC received six applications for national geopark certifications, out of which four were accepted. At the same time, San'in Kaigan geopark was nominated as a GGN membership candidate. During this one-year time, the evaluation process gained a degree of maturity: the management structure and availability of guidance for common people became focal issues for evaluating aspiring geoparks. Regarding the GGN membership nomination, experts were divided in their opinions, and several scientists visited the San'in Kaigan area subsequently to advice this aspiring territory. This pattern repeated itself during the nomination of the Muroto Geopark in 2010.

From 2010, JGN membership application presentations were conducted in the geopark session at the JpGU, which added to the publicity of the geopark nomination process. In 2010, 4 territories were added as national geoparks. 2011 and 2012 saw the successful nomination of 6 and 5 geoparks, respectively. Aspiring geopark numbers steadily rose year by year and in 2013 a total of 10 applications were received. Today, 35 national geoparks (apart from the 8 UNESCO Global Geoparks) exist in Japan. All Japanese geoparks are subjected to reevaluation (since 2012) and several geoparks were asked to improve their performance within a period of 2 years (the so-called Yellow-card system). From 2012, staff members from existing geoparks take part in geopark evaluation, this provides them an opportunity to learn in detail about the benchmarks of geopark activities.

As for the JGN, it organizes an annual gathering of all national geoparks since 2010.[5] As mentioned earlier, workshops and lecture events are also regularly organized; since 2011 geopark workshops are held once or twice every year. Since 2014, the working groups under JGN have added to the debate and reflection process about geopark activities. These working groups have brought together scientists, public officials, and local guides and contents range from research issues to local problems. There is a

[4] The word geopark was independently used in Itoigawa from that time.

[5] The first national geopark convention was held in Itigawa Geopark in 2012.

Geoconservation Working Group that is currently debating issues such as potential pathways to use national level laws such as those pertaining to important natural/cultural properties and national park territories for the protection of geoheritage, and at the same time the working group is pondering over the issue of and how to protect heritage that cannot be brought under these prescribed laws.

8.4 Evaluation System of JGC

JGC members are composed mainly of academicians. Ten members are selected from five academic society, the Geological Society of Japan, the Association of Japanese Geographers, the Volcanological Society of Japan, the Seismological Society of Japan, and the Japan Association for Quaternary Research (these organs nominate representatives to the JGC). Two other members are chosen by JGC with their research expertise in mind. The JGC meets 3 or 4 times a year to discuss on geopark promotion. In order to gain national geopark status, interested territories (applicants) submit application dossier(s) to the JGC in spring (usually in April). These applicants are then asked to present their case at the JpGU in May. This session is also opened to the members of JGN. Field evaluation is subsequently conducted in the summer by three evaluators (at least one evaluator is a JGC member and the other experts are selected from geoparks in Japan by JGC). Based on the report from the evaluators, the JGC decides to approve or reject the application.

The JGC evaluates national geoparks using similar criteria that is followed by UNESCO's IGGP. The two major differences are: while a UNESCO geopark must have at least one geoheritage that is of international value; national geoparks may not possess heritage of such value; and while a UNESCO aspiring geopark must be a de facto global geopark for more than one year prior to the application, national geoparks are not judged at this level (the system allows an area that has just started geopark activity to apply).

In the evaluation of the national geoparks, the following points, which are also important criteria in the evaluation of UNESCO geoparks, are emphasized:

1. Management body: The management body of geopark must be strong enough to coordinate various kinds of activities by stakeholders.
2. Interpretation: Interpretation of geoheritage as well as natural, cultural, and intangible heritage must be simple to understand but at the same time interesting enough to attract visitors and local people. A "geo-story" that explains the geohistory and human history of the area must be created to explain links between various kinds of heritage within geoparks and appreciate them in a holistic way.
3. Conservation: Proper conservation measures must be taken to the important geoheritage.
4. Visibility: Signboard or other promotion material must be established to make geopark visible for visitors.

8.5 Conclusion

In this chapter, I provided an overview of the introduction and maturity of the geopark concept in Japan. Geoparks are a part of an international geoscience program of UNESCO. But at the initial stage, geopark numbers were small and awareness about their activities was low. The initial phase of geopark activities in Japan was marked by the involvement of researchers and organizations who specialized about geology, and the effort to involve local administrative bodies for promotion and management-related matters. The IYPE-related activities (held between 2007 and 2009) became a major catalyst for the diffusion of the geopark concept in the country. Major emphases at that stage were on earth science education and local economic development. While these characteristics are still prominent in Japanese geopark activities today, over the years, geopark activities have matured to incorporate different angles such as the interpretation of heritage, tourism, and protection of heritage. The first global geoparks from Japan were nominated in 2009. Since then several other geoparks have gained the Global Geopark status, and the Japan Geopark Committee (JGC)'s criteria of selecting both global geopark aspirants and national geopark territories have evolved. In addition, the Japanese Geoparks Network (JGN) is now involved in geopark promotion through coordinating between member geoparks and arranging geopark-related events. Through these events and sessions such as those held annually at the Japan Geoscience Union, a perceptible change towards information sharing, debating key issues and raising the evaluation standard can be identified, though several challenges such as effective conservation of geological heritage and achieving sustainable local development remain.

References

Eder FW, Patzak M (2004) Geoparks-geological attractions: a tool for public education, recreation and sustainable economic development. Episodes 27(3):162–164

Iwamatsu A (2007) Naze Ima Jiopaku ka? (Why Geopark is important now?). Chishitsu News 635:8–14 (In Japanese, original title translated)

Iwamatsu A (2008) Nihon Jiopaku-koto Hajime (How did Geopark activity start?) http://www.geocities.jp/f_iwamatsu/retire/geopark_establish.html. Accessed 21 Nov 2016 (In Japanese, original title translated)

McKeever P, Zouros N (2005) Geoparks: celebrating Earth heritage, sustaining local communities. Episodes 28(4):274–278

Taira A, Ohara Y, Wallis S, Ishiwatari A, Iryu Y (2016) Geological evolution of Japan: an overview. In: Moreno T, Wallis S, Kojima T, Gibbons W (eds) The geology of Japan. Geological Society of London, London, pp 1–24

The Shimabara Declaration (2012) The Shimabara Declaration. http://www.geopark.jp/about/pdf/geoparks2012_en.pdf. Accessed 21 Nov 2016

Tsukuda E (2007) Nihon-nimo takusan Jiopaku wo! (Let's establish many Geoparks in Japan!). Chishitsu News 635:6–7 (In Japanese, original title translated)

UNESCO IGGP Global Geopark Guideline (2015) Statutes of the International Geoscience and Geoparks Programme. http://www.unesco.org/new/fileadmin/MULTIMEDIA/HQ/SC/pdf/IGGP_IGCP_UGG_Statutes_Guidelines_EN.pdf. Accessed 23 Oct. 2016

Watanabe M (2007a) Dai 2-kai UNESCO Jiopaku Kaigi Shusseki Hokoku (Report on the Second International UNESCO Conference on Geoparks). Chishitsu News 635, 15–17. (in Japanese, original title translated)

Watanabe, M. (2007b) Sekai Jiopaku ni kamei Suru-niwa? (How to join the Global Geoparks Network?). Chishitsu News 635:42–44 (In Japanese, original title translated)

Watanabe M (2014) Jiopaku no genjo to kadai (Current status and problems of Japanese geoparks). E J GEO 9(1):4–12 (In Japanese, original title translated)

Zouros N (2004) The European Geoparks Network: Geological heritage protection and local development. Episodes 27(3):165–171

9 San'in-Kaigan UNESCO Global Geopark: Geology and Conservation of the Oriental White Stork

Naoki Kikuchi and Kuniyasu Mokudai

Abstract

This chapter provides an introduction of the geological characteristics of the San'in-Kaigan UNESCO Global Geopark and an ongoing scheme of reintroducing wild populations of the Oriental White Stork. Landforms that correspond to a continental margin of Pre-Sea of Japan formation, the formation of the Sea of Japan, and recent sea level rise are found in the coastal areas of this geopark. The area is noted for varied coastal landforms and landscapes of natural beauty, in addition to significant geological sites. The Genbudo Basalt Cave, associated with the formulation of the geomagnetic reversal theory and the Matuyama reversed chron, is a signature geological site; the coastal dunes of Tottori are an important geomorphological feature as well as a major tourist attraction. Recently, a project aiming at the restoration of the Oriental White Stork has succeeded in bringing together a number of local stakeholders for nature restoration. The Oriental White Stork was a familiar bird in seminatural agrarian landscapes, but the species became extinct in the wild after widespread hunting and habitat destruction. The current restoration program aims for "comprehensive nature conservation" by identifying the role of the species as a "marker" of ecological health of landscapes, and in this way, the project represents the linkages between the geological and ecological characteristics of the region.

Keywords

San'in Kaigan • UNESCO Global Geopark • Geological Heritage • Oriental White Stork • Comprehensive Nature Conservation

9.1 Introduction

The San'in-Kaigan Geopark became a member of the Japanese Geoparks Network (JGN) in 2008. It was one of the first geoparks in Japan. Eventually, San'in-Kaigan became a Global Geoparks Network (GGN) member in 2010 (rechristened as UNESCO Global Geopark in 2015). The area falls under the San'in-Kaigan National Park; the east-west extent of 120 km and an area of nearly 2458 km^2 make it one of the larger geoparks in Japan (Matsubara and Niina 2015).

The geopark is known for its distinctive and beautiful rocky coastline and occasional sandy beaches. These landforms also tell us about the formation of the Japanese Islands through the Mesozoic and Cenozoic strata distributed in those locations. San'in-Kaigan is one of the rare places in the Japanese Islands where a relatively large area of such older formations remains protected and visible; thus, the area has both scientific and aesthetic values. Agrarian landscapes are seen away from the coast. These locations are also remarkable for another reason, for being the location of conservation of the Oriental White

N. Kikuchi (✉)
Research Institute for Humanity and Nature, Kyoto, Japan
e-mail: nkikuchi@chikyu.ac.jp

K. Mokudai
Japan Geoservice Inc., Tokyo, Japan
e-mail: kmokudai@gmail.com

A. Chakraborty et al. (eds.), *Natural Heritage of Japan*, Geoheritage, Geoparks and Geotourism,
DOI 10.1007/978-3-319-61896-8_9

Stork (*Ciconia boyciana*) through efforts of establishing a natural population (the species currently survives with the help of artificial habitats and breeding programs). The San'in-Kaigan Geopark therefore is a place where such nature conservation efforts, together with geoconservation attempts to protect the signature sites, and tourism promotion for seaside and hot spring (onsen) destinations are being carried out in order to achieve sustainable development.

9.2 The Geological Heritage of San'in-Kaigan

The three main geological characteristics of this area are:

1. Various igneous and depositional formations from the period of the Sea of Japan formation
2. Varied coastal topography of the Sea of Japan area
3. Post-Sea of Japan formation volcanism and related landforms/landscapes

The geological history of this region goes back to the Cretaceous and Paleogene and was set off by a set of processes that marked the formation of the Sea of Japan. Currently, signatures of that phase are seen from the igneous rocks from that period of history (Sasada et al. 1979). The area also has Ria coast formation and sand dunes created out of crustal deformation and movement of sea levels. Eroded cliffs, sea caves, and abrasion (wave cut) platforms are some of the most conspicuous landscapes formed by long crustal movement and erosion, while sandy beaches, dunes, and sandbars form some of the most conspicuous depositional landscapes (Ota et al. 2004). The inland part is more mountainous (low to moderate elevation), and volcanoes and lava flows from the period of the expansion of the Sea of Japan can be seen in this part (Furuyama 1973, 1990). The climate, ecosystem, and local culture combined with these geological elements give rise to the current agrarian landscape of this area.

9.2.1 Various Igneous and Depositional Formations from the Period of the Sea of Japan Formation

Various igneous and depositional landforms are found in this geopark. The Uradome Coast area (Fig. 9.1) features rhyolitic and granitic formations from 75 to 50 Ma (Cretaceous-Paleogene). These formations are evidence that such igneous

Fig. 9.1 Eroded coastal cliff at San'in Kaigan (Photo by Abhik Chakraborty)

Fig. 9.2 The coastal dunes of Tottori are a major tourist attraction (Photo by Abhik Chakraborty)

activity took place at the continental margin (Sasada et al. 1979), and they date back from a period prior to the opening of the Sea of Japan and the formation of the Japanese Islands as we understand them today. Mammal footprint fossils have been found in 23–5 Ma old sedimentary formations (Yasuno 2003, 2005) in the Takeno area of the Toyooka City; these formations are considered to be the product of rivers depositing alluvium in the continental margin that was beginning to break apart. The Tajimamihonoura of Shin-Onsen cho Town in Hyogo Prefecture was registered as a Place of Scenic Beauty and Natural Monument in 1934; the base of Cretaceous-Paleogene granitic formations is partly overlain with Neogene pyroclast unconformity.[1] The pyroclasts and rhyolitic intrusive formations (dykes) offer evidence of fresh volcanism of the Neogene period (Komuro et al. 2002; Furuyama and Nanao 2004). In the Tajimamihonoura, an outcrop called the Shishino Kuchi (mouth of a mythical beast) (Fig. 9.2) can be seen. This outcrop was formed due to cooling of magma that rose with the expansion of the Sea of Japan (Furuyama and Nanao 2004). As magma rose several times, several magma layers formed, with their interval charred and oxidized in the heat (the reddish bands).

9.2.2 Varied Coastal Topography of the Sea of Japan Area

Most of the geopark's territory is mountainous. Low and moderate level mountains of the Chugoku region dominate the inland part; large plains are not seen in this area. The Ria coast and the gentle coves offer natural locations for ports and fishing villages grew in the northern part of the coastline of Hyogo Prefecture by utilizing these natural landforms. Parts of the coast are eroded; the differential rates of erosion due to geological characteristics form a diverse coastal topography.

The granitic formations of Uradome Coast area are eroded along the cracks that appear on their characteristic columnar joints (Fig. 9.1 above). The rhyolite of Tajimamihonoura is resistant to erosion and forms what is akin to a "residual outcrop." The Tottori Dunes, a major tourist attraction, are located at the mouth of the Sendai River. Dunes are also seen in Kumihama and Kotobikihama areas; these are wind-driven

[1] The unconformity here arises out of nonuniform distribution of age boundaries in the rocks.

(northwesterly circulation in winter) deposits of marine sand along the coast. Currently, vegetation has started sprouting on the dunes, leading to efforts of pruning.

9.2.3 Post-Sea of Japan Formation Volcanism and Related Landforms/Landscapes

Volcanism in this area continued well after the formation of the Sea of Japan. The Genbudo Basalt Cave is probably the most remarkable site that offers evidence of post-Sea of Japan volcanism of this area. This large basaltic cave is located along the Maruyama River; it was formed nearly 1.6 Ma when a large amount of basaltic lava cooled and was subsequently eroded (Sakiyama et al. 2012). The spectacular columnar joints on the cave wall make it one of the signature sites of this geopark. In 1926, Motonori Matuyama discovered that the basalt here has an opposite magnetic alignment compared to the current magnetic poles and formulated his geomagnetic reversal theory based on this (Matuyama 1929). The name of this famous Japanese geologist received its geological permanence in the form of the Matuyama reversed chron (2.59–0.78 Ma). The Genbudo Cave is San'in's main international heritage, for this was the place that provided the first evidence of magnetic reversal in the Quaternary and it was designated as a Natural Monument in 1931 (Sakiyama et al. 2012).

Heritage in a Nutshell: San'in Kaigan UNESCO Global Geopark
Type of Heritage: Geological, Ecological, Cultural
The San'in Kaigan Geopark features landforms predating the Sea of Japan formation nearly 25 Ma and important signature landforms associated with the Sea of Japan formation. The geopark covers three prefectures of Kyoto, Hyogo, and Tottori. Varied coastal landforms, visually beautiful coastal landscapes, moderately elevated mountains, waterfalls, and coastal dunes of Tottori are all important geological/geomorphological assets. The geopark offers several types of tourism, including guided geotours.
The Oriental White Stork restoration program is a remarkable initiative of reintroducing the Oriental White Stork into the wild through participation of citizens, local administration, and scientists. The program is a highlight of the Toyooka City.
Important Facilities: San'in Kaigan Geopark Museum of the Earth and Sea, Hyogo Park of the Oriental White Stork (information available in Japanese only).
Key website: San'in Kaigan UNESCO Global Geopark (http://sanin-geo.jp/en/) (English information available).

9.3 Geological Diversity and Oriental White Stork

The basalt forming the Genbudo Cave has a high erosion resistivity, and the Maruyama River has managed to erode only a very narrow section of the lava flow. On the other hand, the Toyooka valley further upriver is covered by easily erodible granite and forms a basin. The bedrock of the Toyooka basin occurs at 50–70 m depth; a coarse gravelly formation on the bedrock is overlain by marine deposition products (Tanigawa 2009). The occurrence of marine deposits offers evidence that this area was part of a cove nearly 70 ka. In addition due to the occurrence of Jomon Era artifacts in 5 m elevation, it can be assumed that the area was below the sea level (Sakiyama et al. 2012). The Maruyama River itself caused flooding along the river channel, and this formed the conditions for local ecology (Naito et al. 2011). The marshes and subsequent ricefields became an ideal habitat for the Oriental White Stork. However, the species became extinct in the wild in 1971 due to extensive land cover change and development; the last wild bird was found in the Toyooka basin area (Kikuchi 2006). It can be assumed that the geological characteristics leading to local ecology sustained the last wild Oriental White Stork population in Japan. Due to this, the local conservation and reintroduction of the species efforts have a special value as part of the geology–ecology–society narrative. Currently, efforts to reintroduce the species after captive breeding are in progress, supported by the local residents who see the species as of symbolic value to their region. This project is considered a good example of conservation of natural resource and sustainable development (Kikuchi 2006; Sakiyama et al. 2012). A detailed account is provided below (Fig. 9.3).

9.3.1 Oriental White Storks and Humans

The Oriental White Stork was a common bird in Japan as late as in the nineteenth century, but after the Meiji era (1868–912)

Fig. 9.3 Oriental White Stork (*C. boyciana)* Photo by Naoki Kikuchi

modernization, its population dramatically declined due to hunting. An early effort to conserve the species was made in the Tajima area, and it is estimated that 60–100 birds remained by 1935 in that area. However, in 1943 the national forest of this area was logged off and pesticide pollution and agricultural development pressure combined to set in another swift decline of numbers. Efforts to preserve the bird began afresh in 1955 with local administration and residents joining hands and captive breeding began in 1965. However, all this effort did not result in the desired outcome; the last wild bird died in 1971. As there was no colony left in the wild for breeding, the bird became extinct in the wild. The major reasons for the extinction were widespread hunting leading to direct loss of numbers beginning in the late nineteenth century, loss of marshland and pine trees that were essential for feeding and breeding, chemical (pesticide) pollution, and loss of genetic diversity due to loss of viable wild population. All of these reasons are due to human activity; as the bird was a familiar existence in the seminatural agrarian environments, a changing human environment affected the species directly. For the same reason, the Oriental White Stork can be considered as a species that clearly reflects the changes in the nature–society relationship.

Since 1989, the captive breeding program reached a new stage with the successful hatching of an Oriental White Stork. After that, the effort to raise a population in captivity with the eventual aim of reintroduction in the wild began (Kikuchi 2006).

9.3.2 Comprehensive Nature Restoration

In order to reintroduce the Oriental White Stork to the wild, its habitat should be restored first. The Oriental White Stork can be considered a keystone species for seminatural, agrarian environments. The habitat for this species was managed by human activities for several thousand years. However, at present agrarian landscapes face decline due to loss of farmers and abandonment. Therefore, maintenance of the agrarian economy is an important factor, especially the management of rice paddies, and the satoyama landscape is important for habitat restoration for this species. In other words, revitalization of village level economy leading to management of satoyama and paddies is an important pathway for the stork's habitat restoration. Thus, it is necessary to reconstruct the human–nature relationship in a comprehensive manner for this project. As described earlier, the Oriental White Stork is a species that is prone to be influenced by human activities directly. If this relationship is understood correctly, there is a potential that the species can be used as a "marker" for nature restoration in its environment. In the ongoing restoration program, emphasis is placed on the concept that the environment that supports the Oriental White Stork is also the environment that is good for humans. Figure 9.4 shows a symbolic representation of how the Oriental White Stork needs restoration of habitat, restoration of agriculture, revitalization of local culture, networks inside and outside the locality, and revitalization of local economy. All of these aspects are interrelated and need to be addressed together. We refer to this type of combined nature conservation and local development leading to a virtuous cycle of conservation of natural resources and its sustainable use as "comprehensive restoration" (Fig. 9.4).

9.3.3 Change in the Relationship with Satoyama

One of the methods currently employed to aid comprehensive restoration is through the "Oriental White Stork friendly farming method" (Kikuchi 2012). As the Oriental White Stork thrives in a seminatural environment, efforts to increase the diversity of managed natural environment through use of organic farming and promotion of organic farm products are considered a key. Thus, this type of farming can be seen as "valuable" for its contribution to biodiversity, and this cycle can lead to sufficient feeding opportunity for the stork as well as to the restoration of the agrarian economy. Currently, a large number of stakeholders such as the local administration, farmers, agricultural specialists, and farming associations have joined hands to promote such organic farming.

A large typhoon damaged the Tajima area in October 2004, but this event also opened an opportunity for one of the Oriental White Storks (named Hachigoro) to stay temporarily in a waterlogged ricefield in the Toshima area of Toyooka City. This area (marsh) is currently called Hachigoro Marsh; it consists of a part of the ricefield used as an experimental habitat for the stork. A local NPO that promotes the restoration of the Oriental White Stork through citizens action has taken on the work of maintaining this area since 2009. From 2008, regular reports of stork breeding in the pilot site are available. According to Onuma and Yamamoto (2009) and Kikuchi (2012), the value of the stork restoration programs extends beyond Japan; it is estimated to bring in an economic stimulus of 10 Million USD. The link between tourism and local economy can be seen from an example of organically farmed rice being adopted for school meals with the help of a fund generated by tourists.

9.3.4 Progress of the Conservation Program

After 34 years since the species became extinct in the wild, five Oriental White Storks were released from the breeding facility at Toyooka on 24 September 2005. On 31 July 2007, a hatchling taking to the sky from its nest outside the captive breeding grounds was observed, another first in 46 years (Osako 2012).

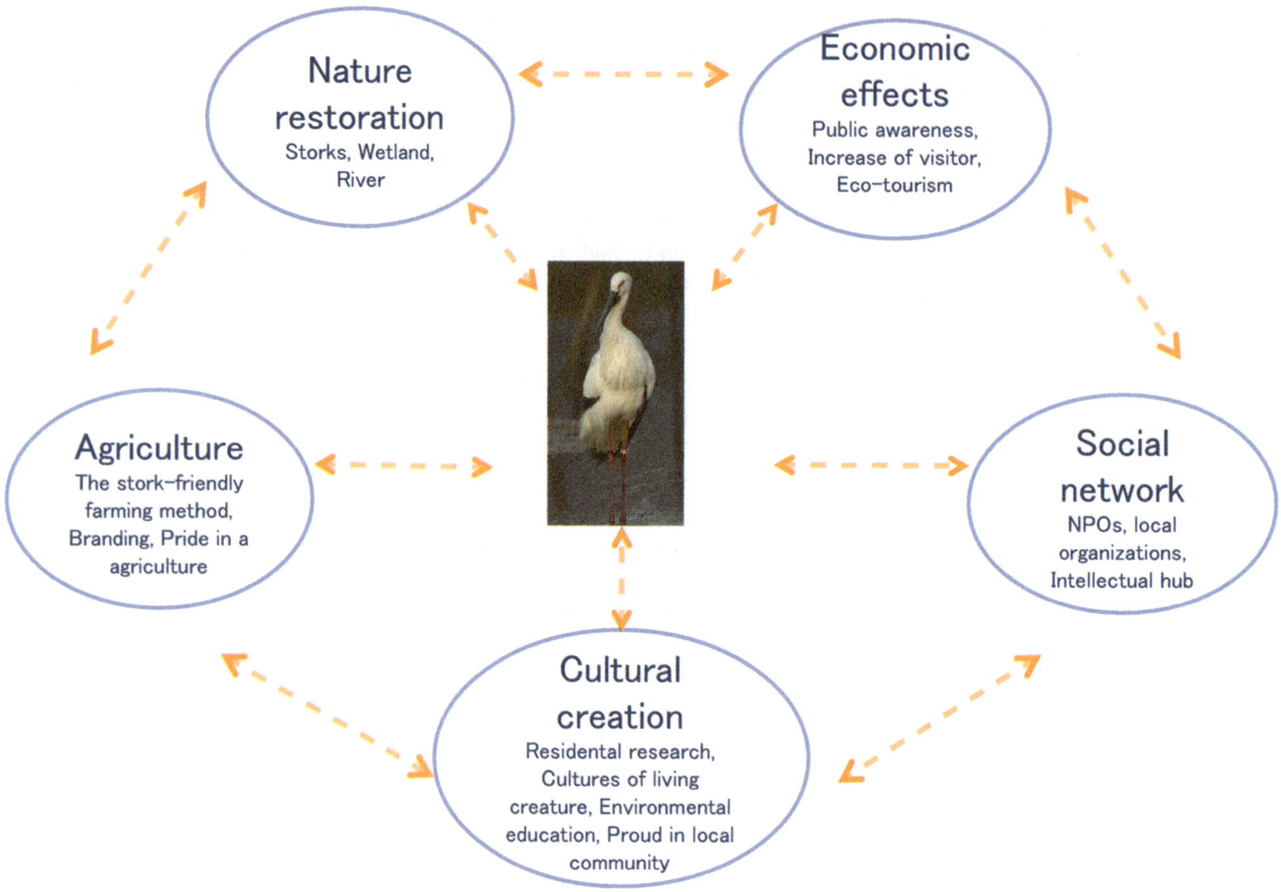

Fig. 9.4 Conceptual representation of comprehensive nature restoration (by Naoki Kikuchi)

Since then, around ten hatchlings are reported to fly out of their nests in the locality every year. In 2011, the third generation of birds released in wild was reported, and currently, an estimated 80 birds remain in the wild (Hypo Park of the Oriental White Stork 2016). The fact that these storks are seen in the San'in-Kaigan Geopark area is an evidence of success of this comprehensive restoration program and a gradual change in the relationship between the local society and the natural environment (Sagawa 2012; Kikuchi 2017). Although most of the birds remain near the Toyooka basin area, some venture outside and cover extensive areas, as far as to the Korean Peninsula. On 3 September 2015, reports of releasing Oriental White Storks to the wild in South Korea became available (Park et al. 2017); in both Japan and South Korea, there is an active involvement of citizens in stork restoration, and an East Asian network has also recently been created.

9.4 Conclusion

The San'in-Kaigan geopark is a geologically diverse area that has landforms and landscapes from the coast to mountains. The coastline has a diverse topography due to the presence of both rocky cliffs and depositional landforms, each of which assume varied appearances. The area is noted for a geological continuity from Cretaceous to the present, leading to a distribution of plutonic, igneous, and sedimentary rocks. The geological diversity of the area created varied natural environments that have a long history of human habitation. The area is also noted for the ongoing program of restoration of the Oriental White Stork to the wild. The Oriental White Stork was once a symbol of this area's natural environment; it is currently adopted as a symbol of comprehensive nature conservation. Currently, an estimated 100 birds thrive in the wild and some of those fly out to cover extensive areas in Japan. However, breeding is currently restricted to the Toyooka basin, with no report of breeding from outside this area available so far. This shows that the physical properties of the Toyooka basin are important for the feeding and breeding patterns of this species. Finally, it should be mentioned that the restoration project itself is an evidence of the combination of geological and ecological characteristics that are unique to this area.

Acknowledgment We thank Abhik Chakraborty for sharing some of the photos of the landscape.

References

Furuyama K (1973) Volcanostratigraphy of the Kannabe Volcano Group. J Geol Soc Jpn 79: 399–406 (In Japanese with English title)

Furuyama K (1990) Kinki Hokubu Kannabe Tansei Kazangunnni Tada Kazan no Hakken (Discovery of 'Tada Volcano' in the Kannabe Monogenetic Volcano Group, North Kinki District). Chikyu-Kagaku (Earth Sci) 44:279–282

Furuyama K, Nanao K (2004) K-Ar Ages of the Teragi Cauldron. Bull Volc Soc Jpn 49:181–187 (In Japanese with English abstract)

Hyogo Park of the Oriental White Stork (2016) Yagai Kotaisu (Outdoor population). http://www.stork.u-hyogo.ac.jp/in_situ/in_situ_ows_num/. (In Japanese). Accesssed 9 May 2017

Kikuchi N (2006) Yomigaeru kounotori:Yaseifukki kara chiiki saisei he (Reviving the oriental White Stork: a reintroduction project for local community sustainability). University of Tokyo Press, Tokyo (In Japanese with English title)

Kikuchi N (2012) A study on re-introduction project of the oriental White Stork and promotion of community development by tourism. Shicchi Kenkyu (Wetland Res) 2:3–14 (In Japanese with English title)

Kikuchi N (2017) "Hottokenai" karano shizen saisei gaku (Natural regeneration starting from the 'can't help caring' sentiment: reintroduction project for the oriental White Stork). Kyoto University Press, Kyoto (In Japanese)

Komuro H, Furuyama K, Shichi R, Hagi N, Yamamoto A (2002) Gravity anomaly on the Teragi Cauldron. Bull Volc Soc Jpn 47:9–15 (In Japanese with English abstract)

Matsubara N, Niina A (2015) San'in Kaigan Geopark. In: Mokudai K, Yuhora K, Niina A (eds) Geoparks of Chubu, Kinki, Chugoku and Shikoku Regions in Japan. Kokon Shoin, pp 96–107 (In Japanese)

Matuyama M (1929) On the direction of magnetization of basalt in the Japan, Tyosen and Manchuria. Proc Jpn Acad 5:203–205

Naito K, Kikuchi N, Ikeda H (2011) Reintroduction of oriental White Storks: ecological restoration and preparation for release into the wild based on the guidelines of IUCN. Jpn J Conserv Ecol 16:181–193 (In Japanese with English abstract)

Onuma A, Yamamoto M (2009) Economic analysis of returning the White Storks to the wild in Toyooka city, Hyogo Prefecture: economic rationale for adopting White Stork friendly farming method and the effect of returning the White Storks to the wild on the local economy. Mita J Econ 102:191–211 (In Japanese with English title)

Osako Y (2012) New progress and reinforcement target in the reintroduction project of the oriental White Stork. Reintroduction 2:21–25 (In Japanese with English title)

Ota Y, Naruse T, Tanaka S, Okada A (eds) (2004) Nihon no Chikei 6: Kinki, Chugoku, Shikou (Regional geomorphology of the Japanese Islands: 6 Geomorphology of Kinki, Chugoku and Shikoku). University of Tokyo Press, Tokyo

Park SR, Yoon J, Ha DS, Cheong SH (2017) A proposal for the habitat restoration of Hwanghae Regions in North Korea through the reintroduction of oriental storks in the Korean Peninsula. Reintroduction 5:53–61

Sagawa S (2012) Process of environmental restoration for reintroduction of oriental White Stork (*Ciconia boyciana*). Reintroduction 2:27–31 (In Japanese with English title).

Sakiyama T, Matsubara N, Mitamura S (2012) Activity of Geopark in San'in Kaigan: relationship between geology and people's lives. J Geol Soc Jpn 118(Supplement):1–20 (In Japanese with English title)

Sasada M, Yamada N, Sakiyama T, Ueda K (1979) Late Cretaceous to Paleogene Igneous Rocks in the Misasa-Okutsu-Yubara Area, Eastern Chugoku District, Southwest Japan. Mem Geol Soc Jpn 17:19–34 (In Japanese with English title)

Tanigawa K (2009) Stratigraphic and sedimentary environment of the recent deposits and Holocene relative sea level changes in the lower Maruyama River Plain, Hyogo Prefecture, Japan. Daiyonki-Kenkyu (Quatern Res) 48:255–270 (In Japanese with English title)

Yasuno T (2003) Early Miocene Pharyngeal teeth (Cyprinidae) fossils from Northwestern San-in and Kyushu, Japan (I). Bull Fukui City Mus Nat Hist 50:1–8 (In Japanese with English title)

Yasuno T (2005) Early Miocene fossil assemblages discovered from Takeno Coast, Tyooka City, Hyogo Prefecture, Central Japan. Bull Fukui City Mus Nat Hist 52:43–65 (In Japanese with English title)

Muroto Geopark: Understanding the Moving Earth

10

Yugo Nakamura and Kazuhiro Yuhora

Abstract

Muroto UNESCO Global Geopark is characterized as a place "Where the ocean and the land meet—the forefront for the birth of new habitable land." This "meeting" takes place over a tectonic plate boundary known as the "Nankai Trough." Tectonic processes have formed the relief and landscape of Cape Muroto: continuous distribution of marine terraces is observed over the sandstone and mudstone base layers in this area. The Muroto Geopark is located at the southern tip of the Muroto Peninsula, southeastern Shikoku Island, and encompasses the administrative district of Muroto City. The geopark covers an area of 248.2 km^2, stretching 18.6 km from east to west and 27 km from north to south, and has a coastline length of 53.3 km. The wedge-shaped cape of Muroto is a distinctive feature of Muroto Geopark. This chapter describes the geological background and the use that Muroto Geopark is making of this by introducing specific aspects of Muroto's Geological Heritage.

Keywords

Accretionary prism • Marine terrace • Tectonic movement • Geopark activity

10.1 Introduction: Where the Ocean and the Land Meet

The earth's surface comprises many tectonic plates, which meet at converging, spreading, and transforming boundaries. Four large plates, the Eurasian Plate, the North American Plate, the Pacific Plate, and the Philippine Sea Plate, converge around the Japanese Islands. In this context, the Muroto Geopark presents geological phenomena typical of plate convergence: submarine trough turbidite beds, an accretionary prism, interplate earthquakes, and tectonic movement. The Nankai Trough (Fig. 10.1) is a 4000-m-deep submarine trench at the Pacific Ocean with a total approximate length of 700 km 80–140 km off southwest Japan, where the Philippine Sea Plate (oceanic crust) moves from southeast to northwest with an approximate velocity of 4–7 cm per year (Seno et al. 1993; Miyazaki and Heki 2001) and subducts beneath the Eurasian Plate (continental crust). Thick submarine sediment deposits are observed at the Nankai Trough, in comparison with other plate converging boundaries, e.g., the Japan Trench (Taira et al. 1992; Hirono et al. 2016).

The Muroto Geopark represents one of the best understood accretionary prisms in the world (Taira et al. 1992). It is located at the southern tip of the Muroto Peninsula, southeastern Shikoku Island, south-western Japan (Fig. 10.1). The geopark is located on the administrative district of Muroto City (Fig. 10.2). Stratified sandstone and mudstone beds observed widely across the cape are a part of the Shimanto Belt (Cretaceous to Neogene, Fig. 10.1). These sandstone and mudstone beds originally formed as "turbidite" deposition layers on the oceanic crust and accretionary prisms

Y. Nakamura (✉)
Muroto Geopark Promotion Committee, 1810-2 Murotomisakicho, Muroto, Kochi 781-7101, Japan
e-mail: nakamura@muroto-geo.jp

K. Yuhora
Tokuyama University, Gakuendai, Shunan, Yamaguchi 745-8566, Japan
e-mail: yuhora@tokuyama-u.ac.jp

A. Chakraborty et al. (eds.), *Natural Heritage of Japan*, Geoheritage, Geoparks and Geotourism,
DOI 10.1007/978-3-319-61896-8_10

Fig. 10.1 Map showing the location of Muroto Global Geopark as well as the Nankai Trough. Source: ETOPO1 Global Relief Model by the National Centers for Environmental Information, NOAA (https://www.ngdc.noaa.gov/mgg/global/relief/), National Land Numerical Information (Administrative Divisions, Emergency Transportation Road, and Rivers) by National Spatial Planning and Regional Policy Bureau, MILT of Japan (http://nlftp.mlit.go.jp/ksj/)

attached to the continental crust. Sandy and muddy sediment, which was once deposited on the river mouth or shallow sea floor, is pushed deeper toward the ocean bottom through turbidity currents when strong storm or earthquakes occur. Eventually, the sediment forms a turbidite bed on the oceanic crust. Off Muroto the turbidite was deformed and dragged into the Nankai Trough with the moving oceanic crust (Fig. 10.3).

The deformed sediments were scraped off the oceanic crust and stacked up on the continental crust during this subduction process, forming an accretionary prism. The development of Japanese Island arc is influenced by the formation of accretionary prism (Taira et al. 1980, 1988, 1992). The concept of accretionary prism is also important to understand formation process of the continental crust (Taira 1990).

Interplate earthquakes are also a common feature of the Nankai Trough. In the historical records, earthquake events of 1707, 1854, and 1946 are documented. Cape Muroto is uplifted with interplate earthquakes (Imamura 1930, Mino 1931, Yoshikawa et al. 1964). On the western coast of Muroto Peninsula, marine terrace surfaces are continuously distributed at several levels of elevation (Fig. 10.4), representing the uplifting trend of Muroto Peninsula (e.g., Watanabe 1932, Yoshikawa et al. 1964, Maemoku 1988, Koike and Machida 2001, Matsu'ura 2015). Elevation and age of the marine terraces imply an uplifting rate about 1.4 mm per year (Maemoku, 2006), which is rapid in comparison with uplifting rates of <1 mm per year (Koike and Machida 2001) at other regions in Japan. Crustal movement caused by plate collision at the Nankai Trough and an east-west directed compressional stress on southwest Japan (Okamura 1990) have produced the acute-angled cape sticking into the Pacific Ocean.

The cape of Muroto, enclosed by the sea on two sides, has suffered natural disasters numerous times in the past. The Nankai Trough causes large earthquakes once every 100–150 years on average (Furumura et al. 2012), and large tsunamis with wave heights of more than 10 m strike the coast every 500–1000 years according to geological studies of paleo-tsunami sediments (Okamura and Matsuoka 2012). Three most recent of such historical tsunamigenic earthquakes occurred in 1707, 1854, and 1946 (Watanabe 1998). In addition to earthquake and tsunami, Muroto is also battered by typhoons that are frequently generated in the northwestern part of the Pacific Ocean during the warm season (Yoshino 1978) and proceed north following the warm Kuroshio current. The destructive typhoons in 1934 ("Muroto typhoon") and 1961 ("Dai-Ni (2nd) Muroto typhoon") caused widespread damage (Kitahara et al. 2012).

A 900-m-depth submarine valley named None Canyon runs from north to south, about 5 km east off Muroto Peninsula (Fig. 10.2). The location of a deep marine canyon off the coast provides access to a wealth of marine resources as well as presenting an opportunity to harvest deep ocean water. Popular catch from around the cape include *Beryx splendens*

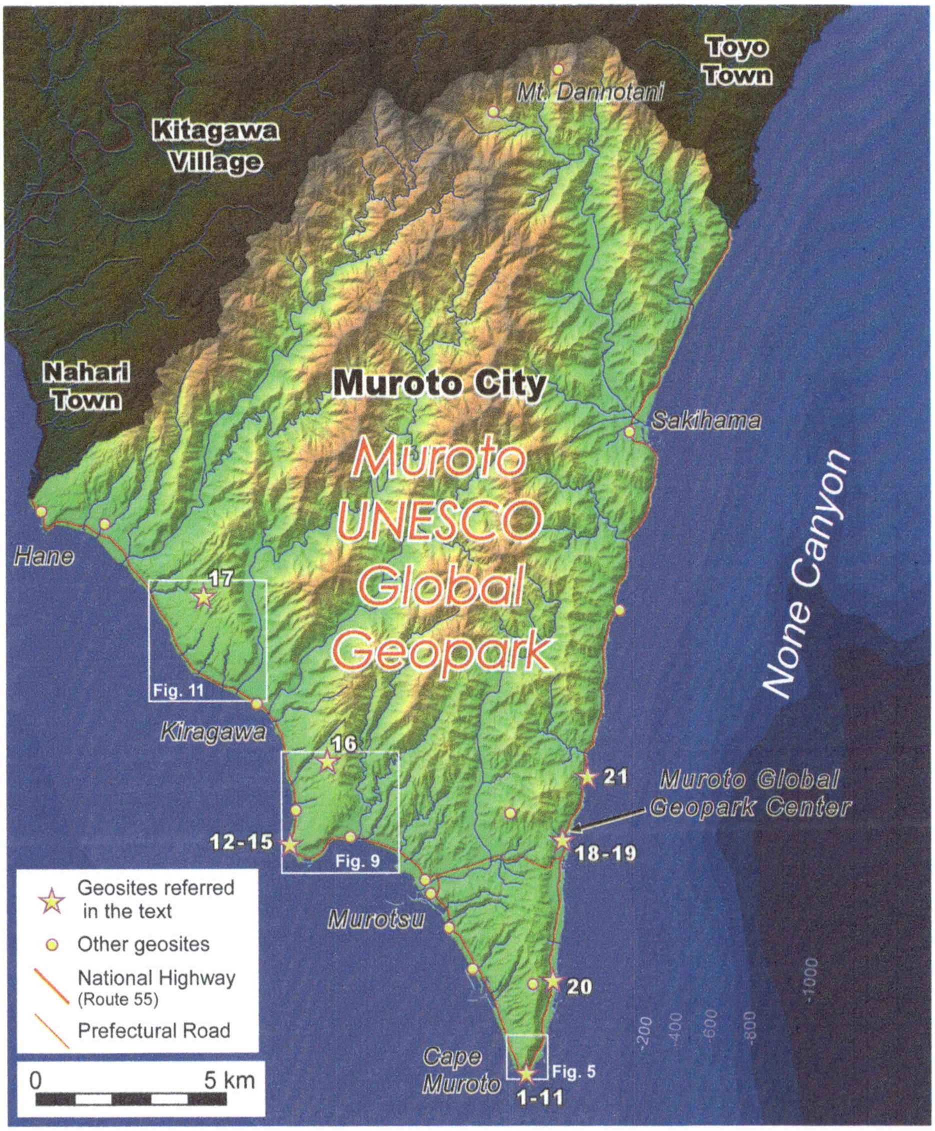

Fig. 10.2 Map of Muroto Global Geopark and distribution of geosites. Source: Base information in Geographic Information System (Map information and Digital Elevation Model 5 and 10 m) by the Geospatial Information Authority of Japan (http://www.gsi.go.jp/kiban/), National Land Numerical Information (Administrative Divisions, Emergency Transportation Road, and Rivers) by National Spatial Planning and Regional Policy Bureau, MILT of Japan (http://nlftp.mlit.go.jp/ksj/), JODC-Expert Grid data for Geography (500 m mesh) by Japan Oceanographic Data Center (JODC), Japan Coast Guard (http://www.jodc.go.jp/data_set/jodc/jegg_intro_j.html)

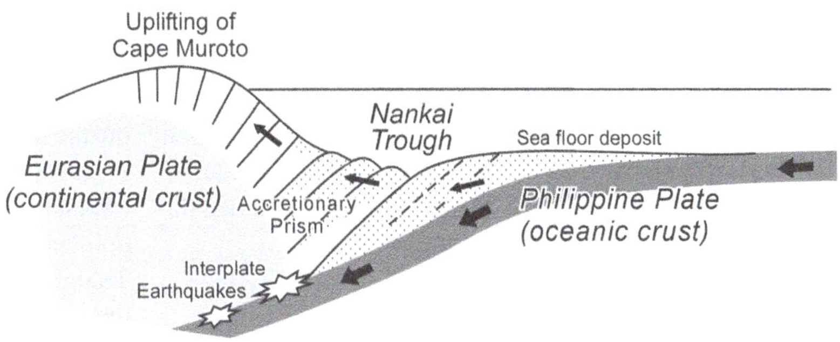

Fig. 10.3 Illustration of the Nankai Trough and relating geological situations (not to scale)

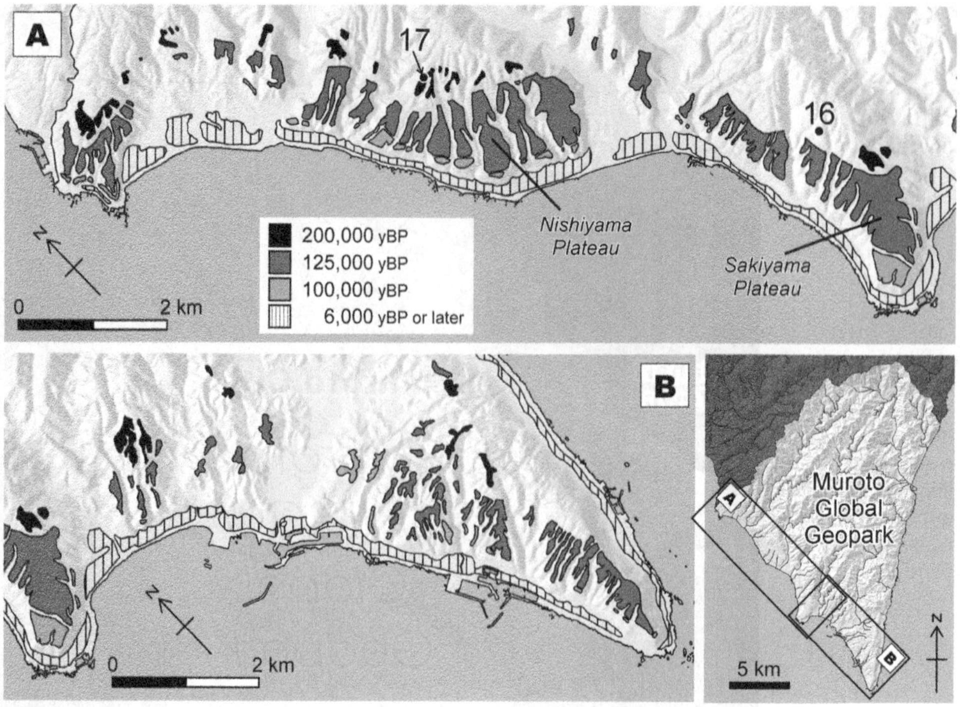

Fig. 10.4 Illustration of the Nankai Trough and relating geological situations (not to scale). Source: Base information in Geographic Information System (Map information and Digital Elevation Model 5 and 10 m) by the Geospatial Information Authority of Japan (http://www.gsi.go.jp/kiban/), National Land Numerical Information (Administrative Divisions, Emergency Transportation Road, and Rivers) by National Spatial Planning and Regional Policy Bureau, MILT of Japan (http://nlftp.mlit.go.jp/ksj/)

(splendid alfonsino), *Katsuwonus pelamis* (bonito), and *Thunnus spp.*(tuna), shells (e.g., *Sulculus diversicolor supertexta* (small abalone)), and seagrass (e.g., *Enteromorpha prolifera* (green laver)).

10.2 Muroto UNESCO Global Geopark: Geological Features, Geosites, and Management Aspects

The Muroto Geopark Promotion Committee was established in June 2008, and the area become a Japanese Geopark in December 2008. Muroto joined the Global Geopark Network (GGN) in September 2011. Along with the adoption of the International Geoscience and Geoparks Programme (IGGP) of UNESCO in 2015, the geopark was accorded the UNESCO Global Geopark status the same year.

10.2.1 Main geosites

Cape Muroto

The Cape Muroto is the most popular geosite in Muroto UNESCO Global Geopark (Fig. 10.5). An outcrop of stratified turbidite bed (Late Oligocene to Early Miocene) (Taira et al. 1980) is observable on the tip of Cape Muroto (Location 1), which is the site of the youngest accretionary prism body on Muroto Peninsula. Its beddings rise almost perpendicular to the horizon (Fig. 10.6), indicating continuous accretionary process and crustal movement. Sand and mud layers in this turbidite must originally have been deposited horizontally; the prism was not only accreted but was uplifted and deformed in the process. Just in front of this outcrop, trace fossils of submarine creatures can be observed (see also the next section on "Gyodo-Kuromi Coast"). To the northeast, forests of the subtropical *Ficus superba* (Location 2), also known as sea fig, and *Quercus phillyraeoides* (Location 3) ("Ubamegashi" in Japanese) are found. The Ubamegashi trees provide the wood for making Binchotan charcoal, which is firmer and better in quality than regular charcoal. These trees typically grow along the coast in temperate or subtropical regions. The raised beach (around location 4) was flattened by wave abrasion and was uplifted by crustal movement (as a result of recent Nankai Trough Earthquakes). Evidences of uplifting can be seen from the former colonies of tube worm *Pomatoleios kraussii* (Locations 5, 6, and 7): this species originally lives in the intertidal zone but uplifting resulted in its former colonies appearing significantly higher than that zone. The basement rock here is gabbro, an igneous rock that is crystallized from magma slowly cooling under the ground. Gabbro cropping out around Locations 8 and 9 (Eboshi-iwa Rock) includes large (about one centimeter) crystals of pyroxenes and feldspars. The magma intruded into the turbidite bed beneath the ocean

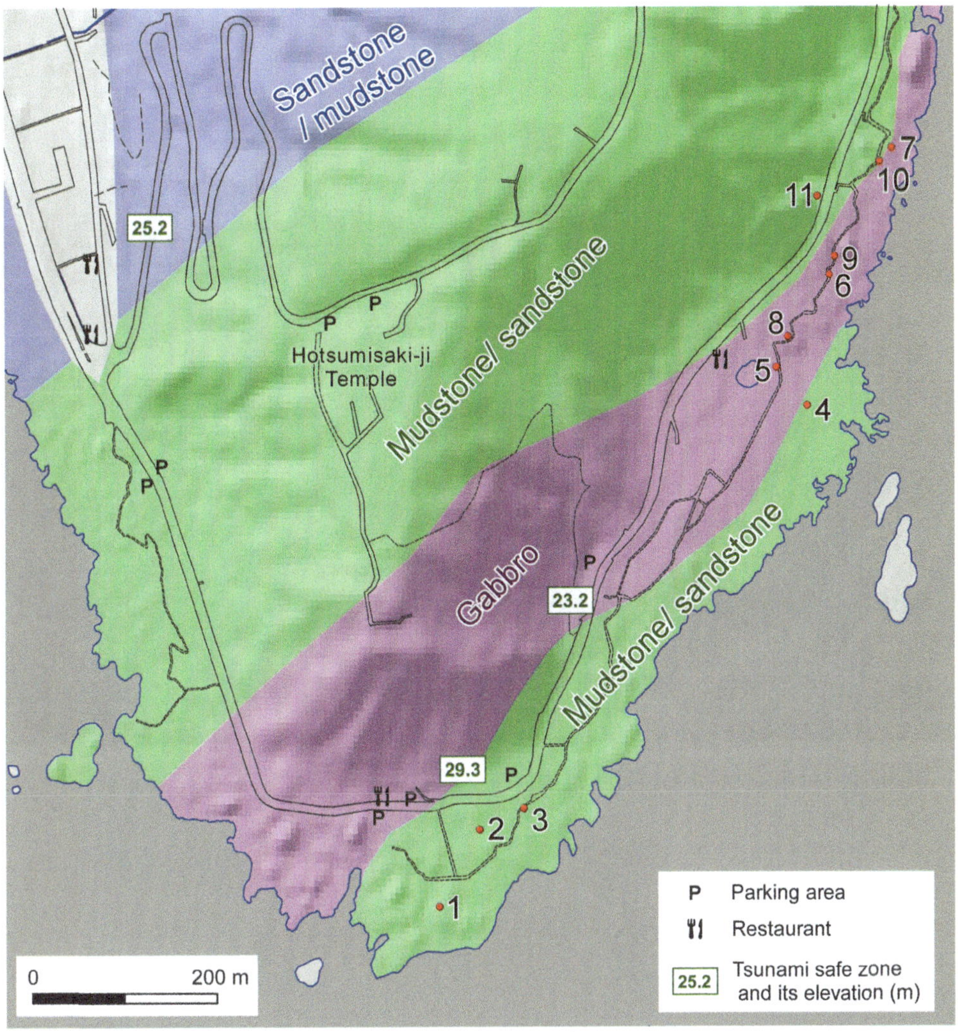

Fig. 10.5 Map of the trail around Cape Muroto and its geology. Source: Base information in Geographic Information System (Map information and Digital Elevation Model 5 and 10 m) by the Geospatial Information Authority of Japan (http://www.gsi.go.jp/kiban/)

bottom at about 14 Ma (Hamamoto and Sakai, 1987). The rock body is small and has a thickness of 200 m (see the geological map in Fig. 10.5). The contact zone between the turbidite and gabbro can be seen at Location 10. Here, the turbidite sandstone was "roasted" by the heat of magma (origin of the gabbro) and was altered to hornfels (a type of metamorphic rock). On the landward side of the national highway (Route 55) (Location 11), an uplifted sea cave, "Mikurodo," can be observed. The entrance was around sea level 3000–5000 years ago, and wave erosion carved the cavity. The cave is also famous as the religious training ground of the historical monk Kukai (AD 774–835).

Traffic: From Kochi Airport to Muroto city center: 63 km (1.5 h by car).

From Nahari Station (Tosa Kuroshio Railway Gomen-Nahari Line) to Muroto city center: 24 km (30 min by car. Bus service is available once an hour from 7 a.m. to 6 p.m.).

From city center to Cape Muroto: 6 km (10–12 min by car/bus).

Gyodo-Kuromi Coast

Typical turbidite layers deposited from Eocene to early Oligocene (30–50 Ma) spread out on the beach of Gyodo-Kuromi Coast (Fig. 10.7). The bedding structure of alternating sandstone and mudstone is deformed at locations (Location 12, in particular) (Fig. 10.8a). This deformation is thought to be the result of a submarine slump (slope movement) deposit (Taira et al. 1992). At Location 13, a 20–30-cm-thick sheet of sandstone dyke penetrates the sandstone/mudstone beddings (Fig. 10.8b). The sandstone dyke was formed after the deposition of turbidite sandstone/mudstone layers on the ocean floor (Taira et al. 1992) and is generally presumed to be the trace of a path along which liquefied sediments run up from the deep to the seafloor surface due to a submarine earthquake. At about 10 m from the end of trail, a wavy surface of sandstone bed can be seen (Location 14). This bedding plane with ripple mark (Fig. 10.8c) represents ancient ripples on the sea floor. On the rock surface at Location 15, there are some long and thin edges (Fig. 10.8d): these are fossilized traces of

Fig. 10.6 Outcrop of turbidite bed (Late Oligocene-Early Miocene) at the southern tip of Cape Muroto (Location 1)

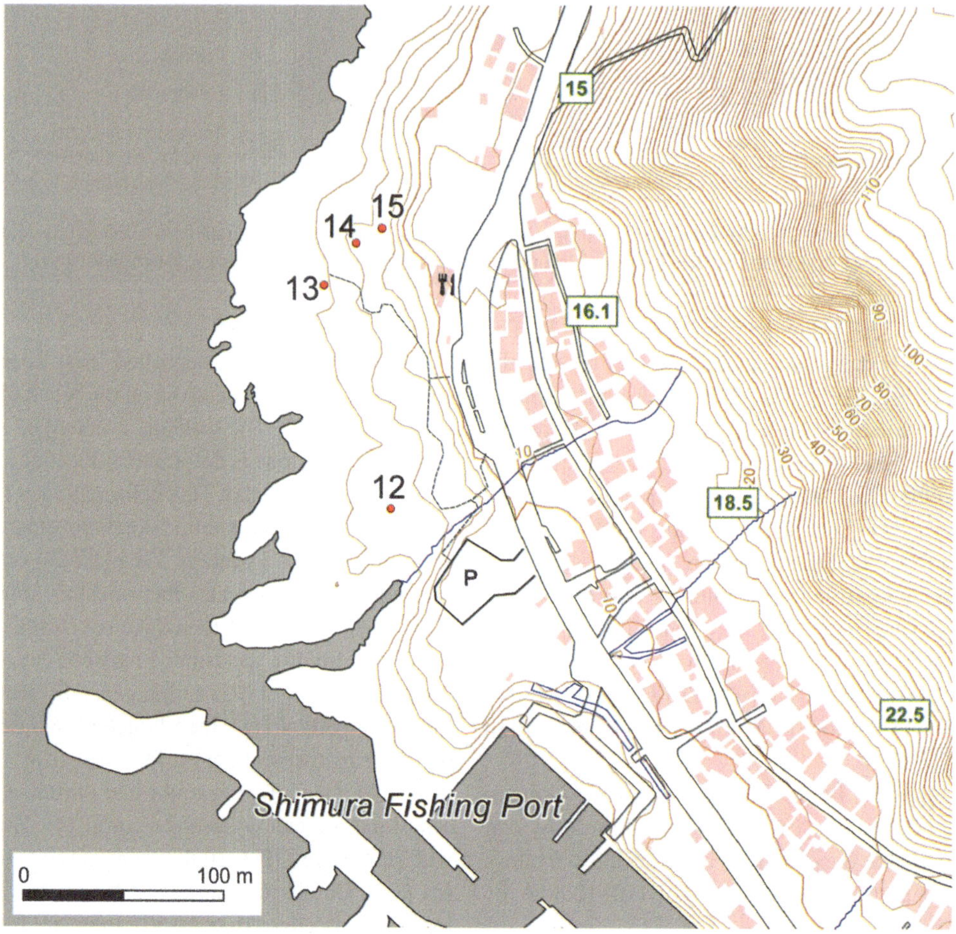

Fig. 10.7 Map of the trail around Gyodo and Kuromi sites. Legends: see Fig. 10.5. Source: Base information in Geographic Information System (Map information and Digital Elevation Model 5 and 10 m) by the Geospatial Information Authority of Japan (http://www.gsi.go.jp/kiban/)

Fig. 10.8 Outcrop of turbidite bed at Gyodo-Kuromi site. (**a**) Slump structure (Location 12), (**b**) sand/clay alteration bed and sand dyke (arrow mark) (Location 13), (**c**) bedding plane of sandstone layer showing ripple mark (Location 14), (**d**) trace fossil in mudstone layer (Location 15)

animal movement (presumed to be bivalves) on the seafloor (Nara and Ikari 2011).
Traffic: From city center: 5 km (10 min by car/bus).

Sakiyama and Nishiyama Plateaus
The whole west coast of Muroto Geopark, from the point of Cape Muroto to Cape Hane, has continuous marine terrace surfaces (Fig. 10.4). Marine terrace is a flat topography formed by wave erosion and deposition on the sea floor that is later uplifted above sea level (or appears after the regression of sea level). In Muroto Geopark, there are at least five generations of marine terraces: about 6000 years old, about 100,000 years old, about 125,000 years old, about 200,000 years old, and about 300,000 years old (Maemoku 1988; Koike and Machida 2001; Matsu'ura 2015). The observation platform of *National MUROTO Youth Outdoor Learning Center* (Location 16, Fig. 10.9) is on a marine terrace that is 300,000 years old; this location provides a vantage point for observing at least three levels of marine terraces (Fig. 10.10). The largest of these terraces, spreading from the Sakiyama Plateau to the Nishiyama Plateau (Fig. 10.4), formed about 125,000 years ago; the elevation of its surface is about 150–180 m. asl on Cape Muroto (southeast direction) and about 100–150 m on Cape Hane (northwest direction). The rate of uplift is about 1.4 mm per year on average; the rate is faster in southeast and slower in northwest.

The raised marine terrace surfaces are used for agriculture in Muroto. Large fields of sweet potato, watermelons, and eggplants are spread out on the main part of the Nishiyama Plateau (Fig. 10.11), a marine terrace that formed 125,000 years ago. The sunny environment and well-drained land of marine terraces are helpful conditions for this type of agriculture. Marine terraces in Muroto do not have thick marine sediment, and occasionally, the eroded rock surface is exposed. On the Nishiyama Plateau, one can see many types of gravel rounded by wave action in a shallow marine environment. An outcrop (Location 17) on the marine terrace of 200,000 years ago shows a layer of well-rounded gravels (Fig. 10.12) whose surfaces are weathered by sub-aerial processes for the last 200,000 years.
Traffic: From city center to National MUROTO Youth Outdoor Learning Center: 8 km (16 min by car). From city center to Nishiyama Plateau: 14 km (30 min by car).

Muroto Global Geopark Center and Surrounding Areas
The Muroto Global Geopark Center (Location 18, 9 a.m. to 5 p.m., open year-round) was opened in 2015. It has exhibitions about geology, geomorphology, local industry, and traditional culture of Muroto Geopark, and it also

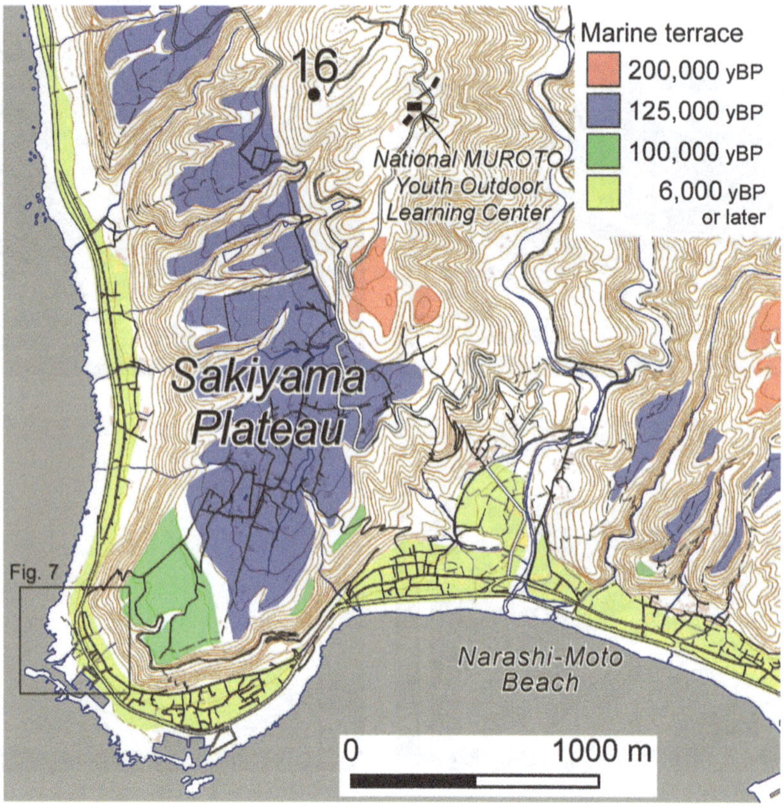

Fig. 10.9 Topographical map showing Sakiyama plateau. See also the caption of Fig. 10.3. Source: Base information in Geographic Information System (Map information and Digital Elevation Model 5 and 10 m) by the Geospatial Information Authority of Japan (http://www.gsi.go.jp/kiban/)

Fig. 10.10 Landscape of marine terrace surfaces, formed 125,000 and 200,000 years ago, from Location 16

features a tourist information counter and a Geo-café/Geo-shop selling local products. The administration office of the Geopark Center is shared by Muroto Geopark Promotion Committee, Muroto City Tourism and Geopark Promotion Division, and Muroto City Tourist Guide Association, and it performs the function as a base facility for administration, research, education, and other geopark activities. In Muroto Geopark, guided tours to three sites (Cape Muroto,

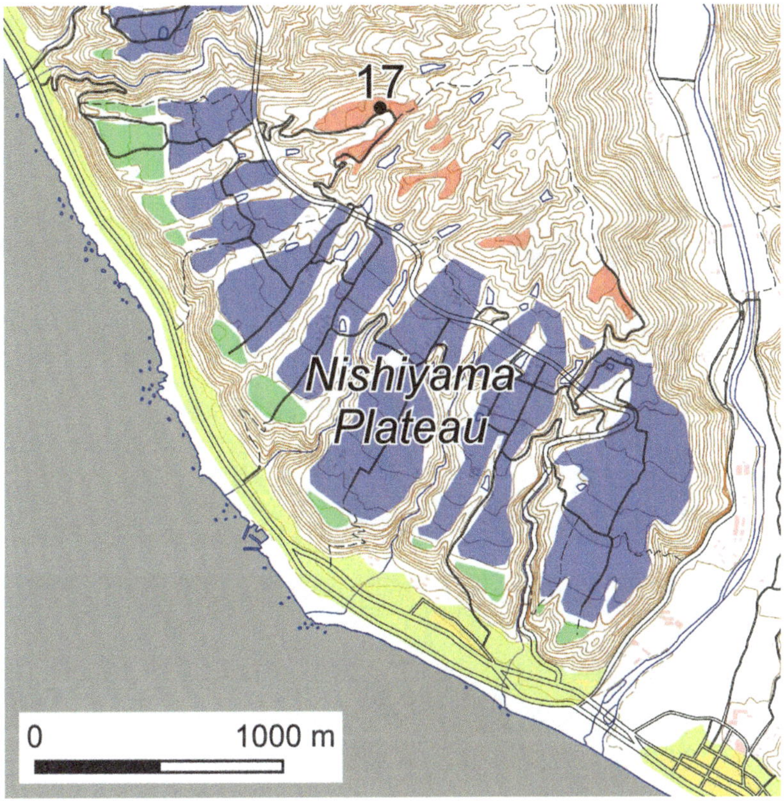

Fig. 10.11 Topographical map showing Nishiyama plateau. Legends: see Fig. 10.9. See also the caption of Fig. 10.3. Source: Base information in Geographic Information System (Map information and Digital Elevation Model 5 and 10 m) by the Geospatial Information Authority of Japan (http://www.gsi.go.jp/kiban/)

Kiragawa, and Sakihama) are regularly operated by guide associations; reservation for these tours can be made at the information counter or via the website of Muroto Global Geopark. The Muroto Geopark Promotion Committee also provides some activity-based programs, such as cycling, sea animal observation at a seashore, and wildlife watch in forests. The deep sea water intake facility "Aqua Farm" (Location 20) and a water processing factory (Location 19) offer exhibitions about deep sea water mechanisms at the East Coast of the Muroto Peninsula. The deep sea water (defined as the water at 200-m-depth or deeper) is taken from 320 to 374 m depth at three facilities on the eastern coast of Muroto Peninsula. The Muroto deep sea water has a low-temperature approx. 9 °C, a high inorganic nutrient (NO_3-N of 12–26 μ mol/dm^3, PO_4-P of 1–2 μ mol/dm^3, and SiO_2-Si of 34–57 μ mol/dm^3) and a small amount of pollutants (TOC of 170 μ mol/dm^3 and marine bacteria of 1–102 CFU/cm^3) (Taniguchi 2001).

In the Hioki Fishing Port (Location 21), about 2 km north from the Geopark Center, visitors can see an example of basaltic "pillow lava." A large boulder (Fig. 10.13 upper), more than 50 m wide, located in the breakwaters, is actually an accumulated body of pillow lava (Fig. 10.13, lower part), formed at 14–15 Ma (Mizoguchi et al. 2009). Igneous rocks distributed around Muroto Peninsula were formed by activities of basaltic magma that intruded into the accretionary prism along plate subduction zone in the Middle Miocene (Mizoguchi et al. 2009).

Traffic: From city center to Muroto Global Geopark Center: 4 km (9 min by car, 21 min by bus).

10.2.2 Other Geopark-Related Activities in Muroto

The Muroto Global Geopark contributes to school education as well as social education. The Kochi Prefectural Muroto High School has "Geopark Study" as a regular curriculum after 2011, the program is supported by Muroto Geopark Promotion Committee staffs. In addition, geosites are utilized by several elementary schools and junior high schools for outdoor learning and study of earth sciences. Geopark activities in Muroto are also conducted in collaboration with local residents. In 2013, five "Promotion Teams" were established in the geopark, one of which is the "Geotourism Promotion Team" that reaches out to a broad range of people with Muroto Geopark Promotion Committee staffs as facilitators. Since July 2013, this team has held regular meetings almost every month, and it has conducted preparatory studies to develop new geopark tours. In

Fig. 10.12 Outcrop of weathered gravels in marine deposit (ca. 200,000 years BP) on Nishiyama Plateau (Location 17)

addition, there are four other promotion units concerning geopark management: network activities, research, and education-conservation-and -disaster mitigation.

Natural and geological heritages of the Muroto Geopark are maintained under the legal framework of the Nature Park Act in the area of Muroto-Anan Kaigan Quasi-National Park and related laws, but systematic efforts for nature conservation were not in place before the beginning of geopark activities in Muroto. Campaigns to gain membership of the Japan Geopark Network and the Global Geopark Network, and subsequent geopark activities, have made the local residents aware of the significance of natural and geological heritage.

In Sakihama region, northeastern part of Muroto, an 18.29-ha tract of old cedar (*Cryptomeria japonica*, thought to be 120–2000 years old) was conserved as a result of geopark activity. This community of large natural cedar trees on Mt. Dannotani was largely unknown even among most residents in Sakihama. However, around the time when the Global Geopark Network membership was decided, the trees gained extensive media attention and the number of visitors to the area grew rapidly. The trees attracted activities by local groups involved in tourism and education, and the forest was designated as "Kyodo no Mori" (or local heritage forest) by the Ministry of Agriculture, Forestry and Fisheries (MAFF). MAFF's "Kyodo no Mori" project aims to recognize the significance of forested areas and to maintain them through community support. The "Sakihama Yakudo Tennen Sugi Kyodo no Mori wo Mamoru Kai" (Association for the Protection of Sakihama Natural Cedar Forest) group was set up with local residents as the core members in 2012. Guided tours in Mt. Dannotani present geo-narrative of how

Fig. 10.13 (*upper*) Basalt boulder in Hioki port (Location 21). Boulder height: approximately 18 m. (*lower*) Pillow lava in the basalt boulder

this community of very large natural cedar trees has interacted with and helped to form the natural environment that sustains it.

Geological and geomorphological heritages in Muroto, including the accretionary prism and marine terraces, have been preserved in good condition and have been subjected to several scientific studies. This is an important fact that has supported Muroto's bid to gain global geopark membership. An agreement on comprehensive collaboration between the Muroto Geopark Promotion Committee, the Kochi Institute for Core Sample Research (Japan Agency for Marine-Earth Science and Technology), and the Research Organization for Regional Alliances of Kochi University of Technology was achieved in 2011; it is expected that this agreement will help to contribute to sustainable development in Muroto Geopark (the agreement between Muroto City and Kochi University was done in 2005 to encourage researches, educations, and regional developments in Muroto City). Such agreements have produced opportunities of learning based on geopark activity such as lecture meetings for local residents on the latest scientific achievements in earthquake and tsunami studies.

Heritage in a Nutshell: Muroto Geopark
Location: Eastern part of Kochi Prefecture, Shikoku Island, Southwest Japan.
Type of Heritage: Geological: Accretionary prisms, turbidite formations, and raised marine terraces associated with oceanic plate subduction at continental-oceanic convergent boundary.

(continued)

Access: From Kochi Airport to Muroto: 63 km (1.5 h by car). From Nahari Station (Tosa Kuroshio Railway Gomen-Nahari Line) to Muroto: 24 km (30 min by car. Bus service is available once an hour from 7 a.m. to 6 p.m.).
What to see: Muroto UNESCO Global Geopark is characterized as a place "Where the ocean and the land meet—the forefront for the birth of new habitable land." Tectonic processes have formed the relief and landscape of Muroto Peninsula. Outcrops of stratified turbidite bed are observable along the coast of Cape Muroto. On the western coast of Muroto Peninsula, marine terrace surfaces are continuously distributed at several levels of elevation, representing crustal uplift and crustal deformation.
Geo-guide: Guided tours of Cape Muroto, Kiragawa Old Street, and Mt. Dannotani (cedar forest) are available. For details, see the website of Muroto UNESCO Global Geopark (http://www.muroto-geo.jp/).
Visitor Center: "Muroto Global Geopark Center": Located in the eastern coast of Muroto peninsula. 4 km from Muroto city center (9 min by car, 21 min by bus). Open all year round, 9 a.m. to 5 p.m.
Useful websites: http://www.muroto-geo.jp/en/

10.3 Issues of Geopark Management

The biggest issue of Muroto Geopark is that most of its geosites are located along the coast and only few are located inland. This means that most of the tourism is concentrated heavily on some spots such as Cape Muroto and Kiragawa Town. A better tourism plan requires allocation of new geo-, eco-, and culture sites in the inland part of Muroto. Geological sites other than accretionary prism, such as fluvial landforms, habitats of rare flora and fauna, and geographical condition of traditional industries, can also add to the diversity. Recently, geomorphological and ecological research on river environment has been started in order to produce new sites and activities in collaboration with local residents. Some geosites are difficult to access (such as the mountain area of Sakihama) and are not frequently visited by tourists; such geosites could be effectively utilized through "geo-trekking tours" or "forest experience tours," At present, geo-guides do not offer inland area tours excepting a tour on Mt. Dannotani in Sakihama region. New tour planning is needed, and cooperation with local people and training programs for geo-guides need to be conducted more actively.

Earthquakes, tsunamis, and typhoons frequently affect Muroto and are important themes of the Muroto Geopark. According to a hazard map published by the Kochi Prefecture in 2012, a large part of coastal plain up to 10 m above sea level in Muroto City would be affected by the predicted next Nankai Trough Earthquake Tsunami. However, practical structure and organization throughout the Muroto Geopark are not well prepared for actual disaster. Establishment of an action plan for disaster prevention/mitigation is indispensable. This issue may also provide a good opportunity to produce new geopark activities such as periodic outdoor experience programs or educational programs for tourists, citizens, and school children. Additionally, organization comprising researchers, citizens, and geopark stuffs should be established for hazard investigations, mitigation, and education.

The fundamental method to enlarge the diversity of geosites is to enlarge the area of geopark itself. For example, if the northern limit of Muroto Geopark is set 40-km further north from the present limit, we can observe the whole Shimanto Belt (Fig. 10.1) after the Cretaceous, the whole of accretionary prism exposed on land, in one geopark, and can also observe topographies affected by both uplift and subsidence along the east coast of Muroto Geopark. This could help highlight the scientific importance, educational potential, and geotourism attraction of this area even more.

10.4 Conclusion

The Muroto UNESCO Global Geopark is a model location for studying accretionary prisms that are distributed in the area due to subduction-related processes at the boundary of the Eurasian Plate (continental crust) and the Philippine Sea Plate (oceanic crust). The accretionary prism comprises of stratified sandstone and mudstone beds deformed and dragged into the Nankai Trough with the moving oceanic crust. On the western coast of Muroto Peninsula, marine terrace surfaces are continuously distributed, representing the uplifting trend. The geopark has numerous geosites that help understand this land formation process, and there are committed guides to take visitors around and explain the geology of the area. The geopark status has also raised awareness about other part of the natural environment such as rare flora, fauna, and landscapes (forests) of visual beauty or ecological importance. There are sites where marine trace fossils and ancient ripple marks of the seafloor are visible on the surface. However, Muroto is located at an region prone to subduction-related earthquakes and tsunami. Several times in the past the area was affected by such geological hazards. The climatic conditions also make the area prone to typhoon damage. Through its local education (school activity) and resident awareness programs, the geopark has been able to raise awareness about such hazards, but more research and dissemination of information remains necessary.

Data Sources for Maps

Figures 10.1, 10.2, 10.4, 10.5, 10.7, 10.9, and 10.11 were drawn by the authors based on the following data sources:

ETOPO1 Global Relief Model by the National Centers for Environmental Information, NOAA (https://www.ngdc.noaa.gov/mgg/global/relief/) (Fig. 10.1)

Base information in Geographic Information System (Map information and Digital Elevation Model 5 m and 10 m) by the Geospatial Information Authority of Japan (http://www.gsi.go.jp/kiban/) (Figs. 10.2, 10.3, 10.5, 10.7, 10.9, and 10.11)

National Land Numerical Information (Administrative Divisions, Emergency Transportation Road, and Rivers) by National Spatial Planning and Regional Policy Bureau, MILT of Japan (http://nlftp.mlit.go.jp/ksj/) (Figs. 10.1, 10.2, and 10.3)

JODC-Expert Grid data for Geography (500m mesh) by Japan Oceanographic Data Center (JODC), Japan Coast Guard (http://www.jodc.go.jp/data_set/jodc/jegg_intro_j.html) (Fig. 10.2)

Photographs were taken by Y. Nakamura

References

Furumura T, Maeda T, Imai K (2012) Tsunami taiseki –butsu kara Kataru Nankai Torafu Kyodai Jishin no Jitsuzo (The real image of the great Nankai trough earthquake derived from Tsunami Deposits. Kagaku) Sci J 82195–200 (In Japanese)

Hamamoto R, Sakai H. (1987) Muroto-oki Hanrei-gantai ni Tomonau Bunsho-hangan no Rb-Sr Nenrei (Rb–Sr Age of Granophyre associated with the Muroto-misaki Gabbroic Complex). Sci Rep Fac Sci Kyushu Univ Geol 15:131–135 (In Japanese with English abstract)

Hirono T, Tsuda K, Tanikawa W, Ampuero J-P, Shibazaki B, Kinoshita M, Mori JJ (2016) Near-trench slip potential of megaquakes evaluated from fault properties and conditions. Sci Rep 6 (1):28184. doi:10.1038/srep28184

Imamura A. (1930) Shikoku Nanbu no Seikyuteki Narabini Manseiteki Chikei Hendo ni Tsuite (Rapid and constant topographical movement in southern Shikoku). Jishin (J Seismol Soc Jpn) 2:357–371 (In Japanese)

Kitahara I, Matsu'ura R Kimura R (eds) (2012) Nihon Rekishi-saigai Jiten (Encyclopedia of historical disasters in Japan). Yoshikawa-kobunkan, 838p (In Japanese)

Koike K. Machida H (eds) (2001) Nihon Kaisei Dankyu Atorasu (Atlas of quaternary marine terraces in the Japanese Islands). University of Tokyo Press, Tokyo (In Japanese)

Maemoku H (1988) Muroto Hanto no Kanshinse Chikaku Hendo (Holocene crustal movement in Muroto Peninsula, Southwest Japan). Chirigaku Hyoron (Geogr Rev Jpn) Ser A 61:747–769 (In Japanese with English abstract)

Maemoku H (2006) Muroto hanto no daiyonki Chikaku Hendo to Jishin Ryuki (Quaternary crustal movement and Coseismic uplift along Muroto Peninsula). Chishitsugaku Zasshi (J Geol Soc Jpn) 112: S17–S26 (In Japanese)

Matsu'ura T (2015) Late Quaternary uplift rate inferred from Marine Terraces, Muroto Peninsula, Southwest Japan: forearc deformation in an oblique subduction zone. Geomorphology 234:133–150

Mino Y (1931) Kochi-shi shikin ni okeru Shinshoku-men no Taihi to Tosa-wan Tohanbu no Kaigan Chikei-shi (Correlation of eroded surfaces around Kochi City and coastal topography of eastern part of Tosa bay). Chigaku Zasshi (J Geogr) 43:256–267 (In Japanese)

Miyazaki S, Heki K (2001) Crustal velocity field of Southwest Japan: subduction and Arc-Arc collision. J Geophys Res Solid Earth 106:4305–4326

Mizoguchi S, Kiminami K, Imaoka T, Kamei A. (2009) Muroto Misaki Chiiki ni okeru Chushin se no Kaiko Kinbo Kasei Katsudo (Miocene near-trench magmatism in the Cape Muroto area, Shikoku, SW Japan). Chishitsugaku Zasshi (J Geol Soc Jpn) 115:17–30 (In Japanese with English abstract)

Nara M, Ikari Y (2011) "Deep-sea Bivalvian Highways": an ethological interpretation of branched Protovirgularia of the Palaeogene Muroto-Hanto Group, Southwestern Japan. Palaeogeogr Palaeoclimatol Palaeoecol 305:250–255

Okamura Y (1990) Shikoku Oki no Kaitei Chishitsu Kozo to Seinan Nihon Gaitai no Daiyonki Chikaku Hendo (Geologic structure of the upper continental slope off Shikoku and quaternary tectonic movement of the outer zone of southwest Japan). Chishitsugaku Zasshi (J Geol Soc Jpn) 96:223–237 (In Japanese with English abstract)

Okamura M, Matsuoka H (2012) Tsunami Taisekibutsu kara Wakaru Nankai Jishin no Kurikaeshi (Nankai Earthquake recurrences from Tsunami sediment). Kagaku (Sci J) 82:182–191 (In Japanese)

Seno T, Stein S, Gripp A (1993) A model for the motion of the Philippine Sea Plate consistent with NUVEL-1 and geological data. J Geophys Res 98(B10):17941–17948

Taira A (1990) Nihon Retto no Tanjo (Birth of Japanese Islands). Iwanami Shoten, Tokyo, 226p

Taira A, Tashiro M, Okamura M, Katto J. (1980) Kochi ken Shimanto tai no Chishitsu to sono Kigen (The geology of the Shimanto Belt in Kochi Prefecture, Shikoku, Japan). In: Taira A, Tashiro M (eds) Shimanto tai Chishitsugaku to Koseibutsugaku (Geology and paleontology of the Shimanto Belt). Rinyakosaikai Press, Kochi, pp 319–389 (In Japanese with English abstract)

Taira A, Katto J, Tashiro M, Okamura M, Kodama K (1988) The Shimanto Belt in Shikoku, Japan-evolution of Cretaceous to Miocene accretionary prism. Mod Geol 12:5–46

Taira A, Byrne T, Ashi J (1992) Photographic atlas of an accretionary prism: geologic structures of the Shimanto Belt, Japan. University of Tokyo Press, Tokyo

Taniguchi M (2001) Muroto kaiyo Shinsui to wa?: Muroto kaiyo Shinsui no Kenkyu to Genjo (What is Muroto deep sewater?: research and current statsu of deep sewater off Muroto) Chishitsu News 565:6–10 (In Japanese)

Watanabe A (1932) The geomorphology of the coastal district of Southeastern Sikoku; a contribution to the knowledge of the recent crustal movements of the area under discussion. Bull Earthquake Res Inst 10:209–234

Watanabe H (1998) Nihon Higai Tsunami Sokan (Comprehensive list of damaging Tsunamis in the Japanese Islands), 2nd edn. University of Tokyo Press, Tokyo

Yoshikawa T, Kaizuka S, and Ota Y (1964) Tosa-wan Hokuto-gan no Kaigan Dankyu to Chikaku Hendo (Mode of crustal movement in the late quaternary on the Southeast Coast of Shikoku, Southwestern Japan). Chirigaku Hyoron (Geogr Rev Jpn) 37:627–648 (In Japanese with English abstract)

Yoshino M, (1978) Kikogaku (Climatology). Taimeido, Osaka

The Interface of Geology, Ecology, and Society: The Case of Aso Volcanic Landscape

11

Shamik Chakraborty

Abstract

Aso, located in central Kyushu, is one of the largest terrestrial caldera volcanoes on Earth. The name Aso actually refers to a group of active volcanoes. The volcanoes of Aso witnessed several very large explosions from 300,000 BP to 90,000 BP, whose ejecta is found over a wide area in Kyushu. The surrounding highlands around the volcano are a major headwater source for the area's rivers such as the Shira River. Some of the largest human-managed grasslands in Japan are found in this area. Aso is a UNESCO Global Geopark for its credentials as a supervolcano and an FAO-designated Globally Important Agricultural Heritage System for its grassland landscape. This chapter introduces the main features of the Aso UNESCO Global Geopark and explains how seminatural grassland landscapes are a result of long-term interaction between the local and regional geology and geomorphology, biotic elements, and human culture. In recent years, the condition of some parts of the grassland has deteriorated; this threatens many ecosystem services (ES) that they produce. This chapter evokes the concept of traditional socio-ecological landscapes that are known as "satoyama" in Japan, as an important management issue for this remarkable volcanic geoheritage.

Keywords

Cultural landscapes • Ecosystem services • Aso volcanic area • Japan

11.1 Introduction: Structure and Characteristics of the Aso Landforms

Japan's particular geology of an active island chain sitting in a location influenced by the interaction of four significant plates (the Eurasian, the North American, the Philippine Sea, and the Pacific) makes its geological and landscape diversity unique. Many of Japan's signature landforms and landscapes are found in active volcanic areas. The case of the Aso volcanic system (Fig. 11.1), discussed in this chapter, shows both the complexity of landscapes and links between geology, biodiversity, and culture. The Aso area has small geological units such as the Tateno gorge (formed by alternative actions of fault systems and volcanic lakes), which adds to the internal diversity, whereas larger units such as the lava plateau represent important elements on the global scale. Several centuries ago, Buddhist monks used the area for pilgrimage to the "hi-no-yama" or mountain of fire, which was the name of Aso at that time. Long stretch of plains with monk quarters have been found west of Mt. Naka dating to Kamakura (1185–1333) to Muromachi (1392–1573) periods (Aso Geopark 2012c). Also, signs of ancient people's usage of the iron-rich yellow ochre deposits in decoration and artworks have been found in the region and the Aso obsidian was used for prehistoric artifacts

S. Chakraborty (✉)
Institute for the Advanced Study of Sustainability, United Nations University, Tokyo and Integrated Research System for Sustainability Science (IR3S), The University of Tokyo, Tokyo, Japan
e-mail: tsubakurodake2003@gmail.com

A. Chakraborty et al. (eds.), *Natural Heritage of Japan*, Geoheritage, Geoparks and Geotourism,
DOI 10.1007/978-3-319-61896-8_11

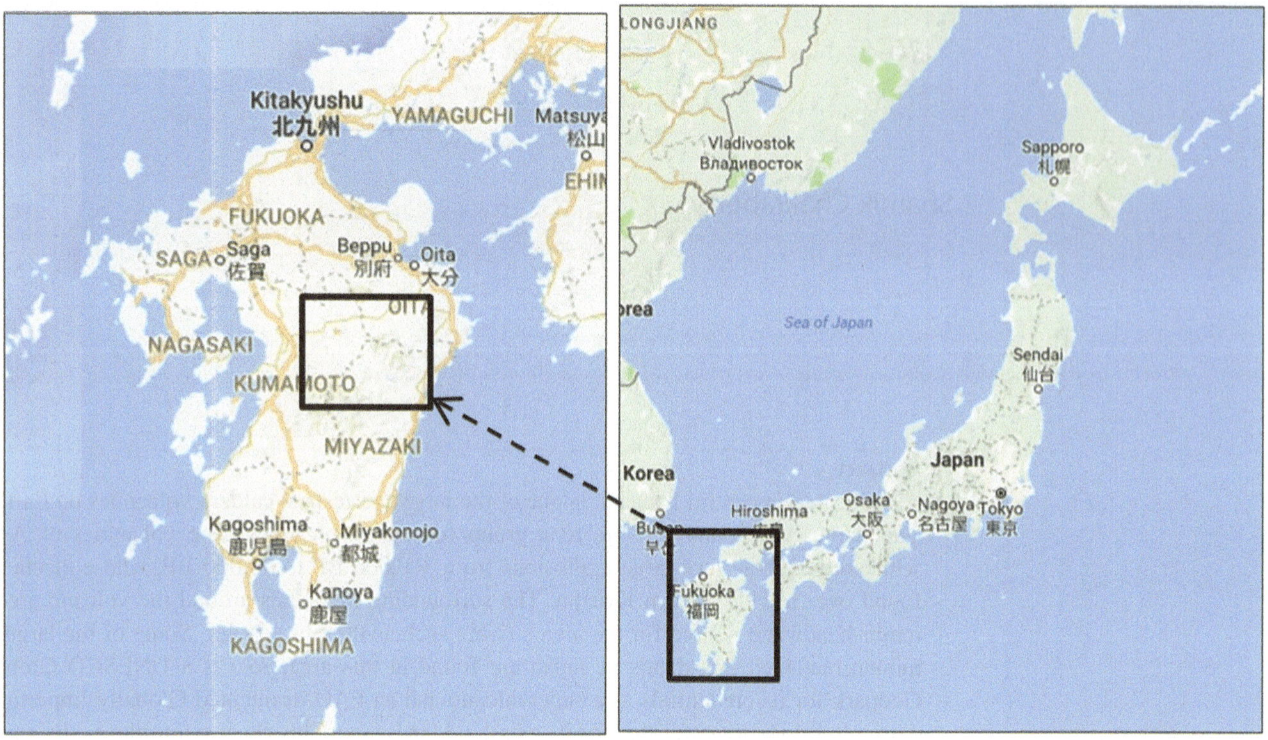

Fig. 11.1 Location of the Aso volcanic area in Kyushu. Source: Google Map

(Nakai et al. nd). The beautiful scenery of the rolling plateau and ancient crater complexes covered by grasslands is a major tourist attraction.

11.2 Geology and Geomorphology

11.2.1 The Aso Area

The landforms of Aso are products of volcanic activities due to the subduction of the Philippine Sea plate beneath the Eurasian plate (Chapman et al. 2009). The resultant arc type volcanism has created the Aso landscape. The Aso caldera is situated in the center of the geotectonic lines with active fault zones marked by the red patches in Fig. 11.2 below:

The most prominent landform in the area is the Aso caldera with five "peaks" at the central part. With its area of 350 km^2 (North-South diameter of 25 km and East-West diameter of 18 km), it is the second largest caldera in Japan (Miyoshi et al. 2013) after the Kussharo caldera in north-eastern Hokkaido. A group of five peaks, Mt. Taka, Mt. Naka, Mt. Neko, Mt. Kijima, and Mt. Eboshi, are located inside this caldera. These newer landforms are the result of post-caldera-forming eruptions (Miyabuchi 2009) (Fig. 11.3), and they form a spectacular "volcano within a volcano" landscape (Fig. 11.4). Before the caldera-forming eruptions, Aso was characterized by thick lava flow and plateau landforms (Ono and Watanabe 1985) (Fig. 11.5). Large-scale caldera-forming eruptions are among the most catastrophic geological events (Miyoshi et al. 2013). The Aso system of calderas and volcanic cones are results of multiple eruptions with pyroclastic flows, four of which are related to major caldera formation phases. These are Aso-1 (300,000 years ago), Aso-2 (140,000 years ago), Aso-3 (120,000 years ago), and Aso-4 (90,000 years ago). Among these Aso-4 eruption was the strongest. The scale of the eruption was so large that the fallout tephra and volcanic ash were deposited inside the sediments of Sea of Japan (Machida and Arai 1983).

As a result of these catastrophic geological events, very large and extensive pyroclastic flows took place, giving rise to the distinctive Aso rim area, also known as the Aso lava plateau. The lava plateau forms the perimeter of the Aso caldera. Large eruptions spewed out so much magma (according to Aso UNESCO Global Geopark (2012a, b, c, d), Aso-4 alone ejected about 600 km^3 of volcanic products) that in order to maintain isostatic equilibrium, some part of the continental crust had to crumple inward. This zone of inward movement created the caldera. In its earlier times, the caldera was dotted with active volcanic conduits. Later, after cessation of in-caldera volcanic activities, rainwater and groundwater filled the area of the old craters to form volcanic lakes and grasslands (Kusasenri, one of the most popular tourist attractions at Aso, is an old volcanic crater that

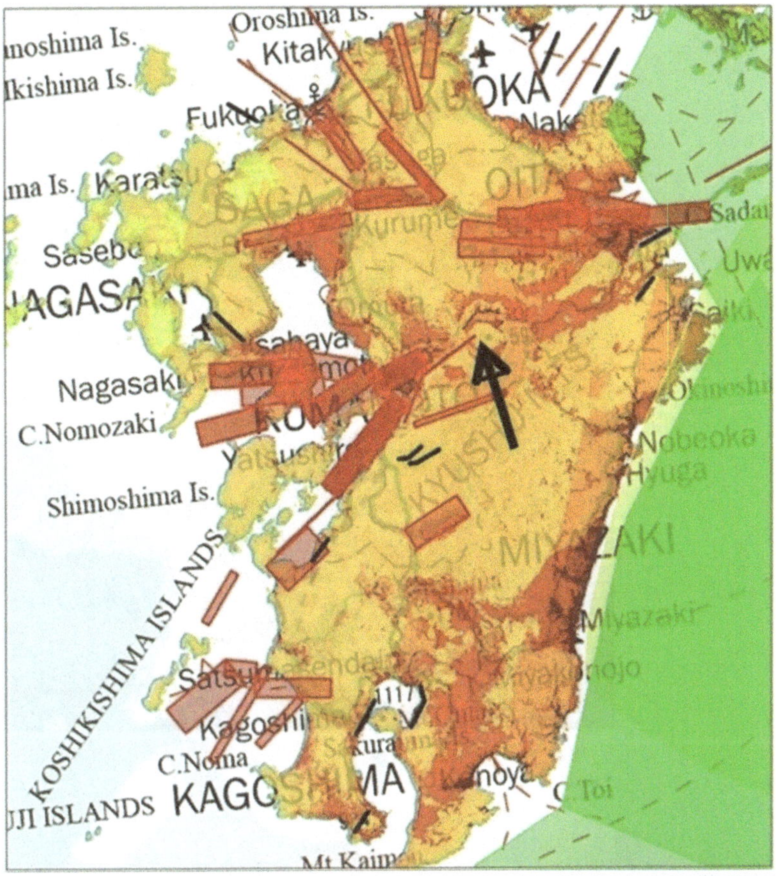

Fig. 11.2 *Above*: Major Active fault zones (*red patches*), fault lines (*black lines*), and subduction zones (*green patches*) of Kyushu (Source: JSHIS 2011, accessed from http://www.j-shis.go.jp, reproduced with permission). The *black arrow* shows the location of Aso caldera. *Below*: The Aso Caldera (Source: Wikipedia, NASA Shuttle Radar Topography Mission)

underwent substantial biophysical weathering (Fig. 11.4). Miyoshi et al. (2013) point out that the existence of a very large magmatic system gives rise to large Aso-scale eruptions. Such magmatic systems are keys to understand active caldera volcanoes. Aso's caldera-forming magma chamber also created a long-distance lateral flow of magma that gave rise to Takayubaru, Togawa, and Tamaraigawa lava units (ibid). It is speculated that the strike slip faults of Oita Kumamoto tectonic line may have caused this long-distance lateral flow of magma (Watanabe and Ono

Fig. 11.3 Mt. Naka active crater and the tephra landscape known as "*sunasenri*" nearest to Mt. Naka. Photographs by the author. Oct 3, 2009 (*above*) and July 25, 2011 (*below*)

1969). The rocks produced from Aso show particularly high concentrations of potassium, and silicic rocks were formed mainly after the Aso-1 to Aso-4 eruptions (Ono and Watanabe 1985).

Other than the volcanic landforms of the caldera, several other volcanic features hold keys to the formation event of the Aso landscape. The Tateno gorge exhibits the alternative works of faults systems that caused several volcanic lakes to

Fig. 11.4 Kusasenri with two old craters (shown with *white arrows*) and Mt. Eboshi. Photograph by the author. Taken Aug 8 2014

form and disappear over time. Aso's iron-rich compounds of red and yellow soil (limonite) were used for artworks in ancient times. These red and yellow ochre deposits are found in the Asodani valley, and they indicate the presence of lacustrine deposits of old lakebeds to the northwest of the caldera. Huge pyroclastic flow deposits can be seen near Oguni town to the north of the caldera, and these are another evidence of the scale of caldera-forming eruptions. An iconic feature of the Aso landscape is the "Komezuka," a small monogenetic volcano made up of basaltic scoria and covered by seminatural grasslands.

The geomorphological (surface landform) features of Aso are intricately linked with geology, and this provides a number of benefits for the local society and ecology. The black soil of the region is formed by volcanic ash depositions. It has good organic matter content and is easy to plow. Good permeability and water holding property makes it an important resource for grassland ecosystems. This in turn helps agriculture: vegetable gardens, fruit orchards, and stockbreeding are prominent economic activities in this area. The porous soil structure let water infiltrate easily, forming a rich groundwater table. Six rivers (Ono, Gokase, Midori, Shira, Kikuchi, and Chikugo) and about 1500 freshwater springs (Takahashi 2014) are related to the Aso landform and landscape. Kumamoto City, the capital of Kumamoto Prefecture with a population of 731,815 (Kumamoto City 2012), gets its drinking as well as agricultural water from the Shira River, which originates from surface water runoff and groundwater recharge of the Aso caldera. The Miyaji, Yakuinbaru, Minami-Aso, and Takamori areas have a number of springs. The bountiful freshwater resources make agriculture easy for the villages that dot the caldera floor, giving considerable economic benefits to the local community. The caldera walls house talus slopes and alluvial fans where steepness of the surface suddenly decreases (structure) and causes deposition of alluvium (function). The walls of the Aso caldera define the catchment area of the River Shira. The rivers Shira and Midori act as major regional geomorphological agents, carrying alluvium generated by weathering and mass wasting. These rivers eventually connect the landscapes of the caldera to the coastal ecosystems, and they sustain a large mudflat area (Shira River estuary) of about 600 ha at the coast of the Ariake Sea. Mudflats are formed due to combined action of salt- and freshwater and tides. Silt and fine-grained alluvium brought by the rivers are main materials that are deposited in a process unfolding over thousands of years. The mudflat of the Shira River estuary is an important migratory bird area in Japan. About 20 species of waders and ducks live and breed in this area, feeding on flora and fauna that are specially adapted to mudflat ecosystems (Wild Bird Society of Japan 2009). Thus, landform and landscape features and their supporting ecosystem services give rise to a number of important processes that encompass a large area.

11.2.2 The Kuju Area

Adjacent to Aso, the Kuju system of volcanoes with five peaks—the Mt. Kuju, Mt. Naka, Mt. Mimata, Mt. Taisen, Mt. Hossho, Mt. Waita, Mt. Kuro—and plateaus formed by pyroclastic deposits is a major part of the Aso-Kuju National Park (727 km^2) (Fig. 11.6). The area is also very popular with hikers in summer. Kuju is one of the most active volcanoes in Japan. Mt. Io (pronounced ee-oh) spews SO_2 in the atmosphere constantly. The majority of the peaks are volcanic domes with some stratovolcanoes. The Kuju system has erupted 8 times since 1662 (AIST, GSJ 2017). Kamata and Kobayashi (1997) have shown through study of Kuju's tephra formations that the volcano has about a 1000-year eruption interval for the last 5000 years. Kuju is a new volcano as well, as 70% of the volcanic landform appeared in the last 15,000 years. The most recent eruption was in 1995 from Mt. Hossho. This eruption was the result of

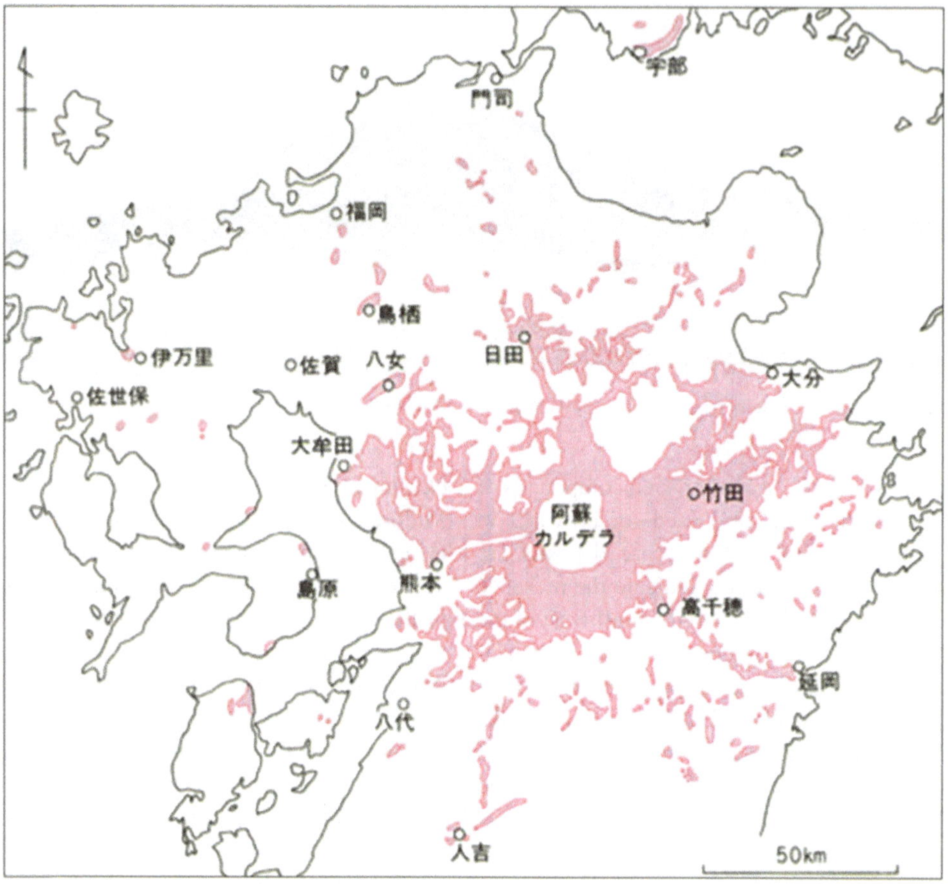

Fig. 11.5 Extent of the pyroclastic flow deposits of caldera-forming eruptions of Aso. Source: Distribution of Aso Pyroclastic Flow Active volcanoes of Japan-Aso volcano (Ono and Watanabe) 1985, https://gbank.gsj.jp/volcano/Act_Vol/aso/text/eng/exp04-1e.html# Geological Survey of Japan, AIST. Reproduced with permission, English caption by author

Fig. 11.6 Mt. Mimata with active Mt. Io with gaseous eruptions seen in the extreme *left*. Photograph by the author. Sep 14 2014

increased groundwater pressure (phreatic). One of the biggest eruptions of Kuju can be found in the deposition of A2 ash of pyroxene-hornblende andesite (12,300 BP–9927 BP), but it cannot be singled out as the cause of A2 ash; three other volcanoes, Mimata, Hossho, or Yuzawa, might have been the main cause. The most recent of the major (due to magmatic activity) eruptions was the lava dome building eruption of Mt. Kurotake about 1500 years ago. Based on its geological history, Kuju is listed as one of the "hazardous volcanoes" in Japan (ibid).

11.3 Natural Heritage Management at Aso

11.3.1 International Heritage Labels in the Area

The Aso-Kuju area was designated a national park as early as in 1934[1] (it was one of the first national parks in Japan) due to its complex and important geological landscape, and the area around Aso was recognized as a global geopark in 2014 (the next year, it became an official UNESCO global geopark). Japan now has a total of eight such global geoparks: Itoigawa, Unzen, Toya caldera, and Usu volcano; San-in seashore, Muroto, Oki, and Mount Apoi are the other members. The volcanic landscape of Aso and its specific landforms were recognized as an important heritage under the International Geoscience and Geopark Program (IGGP), which recognizes geographical areas that possess internationally valuable geological heritage. Geoparks are not only about geology, and they must feature adequate conservation efforts, educational importance, and sustainable development plans.

In 2013, the Aso landscape was designated as a Globally Important Agricultural Heritage System (GIAHS) under UNFAO (FAO 2017). There are increased activities to attract visitors to experience the grand beauty of the caldera with the ash and gas spewing Mt. Naka crater still visibly active, along with the cultural significance such as the managed grasslands that provide important economic resources. Geotourism in Aso Global Geopark follows three geological themes: (1) old volcanism through the present landscapes at Aso Geopark, (2) present volcanism through Nakadake crater activities, and (3) the relationship between volcano and people mainly in the old caldera and the rim area (Aso Geopark 2012a, b, c, d). However, there are many challenges for the sustainable management of the Aso area. The island of Kyushu is one of the most human altered main islands of Japan, and Aso's geoheritage comes under pressure from urbanization, unsustainable land use practices, and construction of dams and roadways. It is important that the geopark looks beyond the tourism perspective and stresses more on the structure, function, and process of the Aso landscapes with their connectivity between the landscapes and their changes; geoparks must explain the abiotic linkages with the biotic communities and promote the understanding of complex components of ecosystems. This role of geoparks is seen as an important part of a comprehensive framework for management of landscapes (Erikstad 2013). An ecosystem service framework, which has recently gained attention in the fields of ecosystem science and sustainability science, may be a viable starting point for grounding this approach, a fact also pointed out by Gray (2013).

11.3.2 Aso Geopark: A Brief Outline

The main attractions of the Aso UNESCO Global Geopark are volcanic landforms and landscapes created out of volcanism related processes. The three main themes of the geopark are (1) the caldera-forming eruptions (supervolcano eruptions), (2) secondary eruptions, mainly those at the Mt. Naka crater that symbolizes highly visible and accessible active volcanism, and (3) by-products of volcanism such as springwater, hot springs, and grasslands (Aso UNESCO Global Geopark 2012a, b, c, d). A crater complex that includes the restless Mt. Naka and older, inactive Sunasenri and Kusasenri and the monogenetic Komezuka volcano form the core of the caldera area geosites. The Shira and Kuro river watersheds support lowland settlements inside the caldera; large columnar joint formations and intrusive rock (dyke) formations of Rakudayama volcano are other notable attractions (Ibid). Notable among outer rim area geosites are the Soyokyo Gorge where towering cliffs of ignimbrite are exposed in a narrow 15 km incised valley of the Gokase River; a spectacular ignimbrite riverbed observable at the Yusuikyo Gorge; and the Rakanyama geosite where pre-Aso stage andesitic rocks of curious formations are seen.

The Mt. Naka and the Kusasenri are probably the best-known geosites. Their locations close to the Aso Museum make them easily accessible. The restless Nakadake makes approach difficult time to time with minor phreatic eruptions, but in overall terms the volcanic activity is gentle. However, the recent earthquakes damaged many geosites; a detailed condition of geosite condition and access is available at the Geopark's homepage (http://www.aso-geopark.jp/en/info/160726.html).

Due to the large internal diversity of geosites, the geopark can be enjoyed through a variety of treks, tours, or casual visits. The geopark website introduces tours on themes like volcanic landform observation at Aso City, stream walking at Minami-Oguni Town, and forest walking at Minami Aso village. Visitors can also soak in the hot springs and enjoy the artesian spring (cold spring water) landscape of the Minami Aso village. The area offers a rich variety of experiences and has a high geotourism potential.

[1] Aso-Kuju was designated as a national park in Dec 1934; before that, Seto Naikai, Unzen, and Kirishima became the first NPs in March 1934.

Heritage in a Nutshell: Aso
Location: Kumamoto Prefecture, Central Kyushu.
Access: About 2 h flight from Tokyo to Kumamoto Airport, take Airport bus to Beppu Ekimae Hon Machi and get off at Aso Station (about 1 h ride). Useful transport inside the Aso area include Oguni Gurutto Bus (Tel: 0967-46-3121), Aso Boy Rail (Tel: 050-3786-3489), Minami Aso Yurutto Bus (Tel: 0967-67-2230), Yuzuke gou Rail (Tel: 0967-62-0058), and Aso Teiki Kankou Bus (Tel: 0967-34-0211)
Type of Heritage: Active volcano complex with second largest caldera in Japan, and largest human managed grassland landscape.
What to see: The spectacular Nakadake volcano and crater complexes with active volcanism of gaseous eruptions. One can take the Aso-san Ropeway from Aso-san Nishi station to see the crater from close range (please note that there are times the ropeway is closed due to Nakadake eruptions). Other volcanic landscapes such as scoria dust, volcanic bombs, cinders, and unique plants such as Japanese Knotweeds that can survive in the harsh volcanic environments make some characteristic features of the landscape. Old and extinct craters at Kusasenri with grasslands and marshes produce iconic vista. Seeing the Nakadake eruptions from Kusasenri makes it possible to experience active and old volcanism together. The Daikanbo area offers a spectacular panoramic view of the central volcanic cones, old caldera, and grassland landscapes. Many other volcanic features like Takayubaru plateau, Togawa and Tamarai River lava units, and Ikenokubo maar are associated with volcanic landscapes that represent the Aso's past eruption history. Other volcanic cones include Takadake, Nekodake, and Eboshidake, with Takanoobane dake (lava dome) and Komezuka (scoria cone). The picturesque Nabe ga taki (waterfall), located in the Oguni town, is another attraction that can be reached through about one-hour drive from Aso station. The primary forest of Kitamuki mountain (Aso Kitamuki Valley Primary Forest) in Minami Aso village (30 min drive from Aso station) is another ecological asset with old-growth evergreen broadleaved (laurel) forests. The area offers good opportunity for bird watching and butterfly viewing in the grasslands and surrounding areas.
Hot Springs: Numerous hot springs characterize the active volcanic area of Aso: Uchinomaki, Akamizu, Jigoku, Tarutama, Yunotani, Tochinoki, Teno, and Katasumi hot springs, to name a few. Different hot springs are also characterized by different mineral contents, as groundwater seeps through different properties of rocks.
Aso Visitor Center: Located in Kumamoto Prefecture, Aso City's Kurokawa area. Opening hours are from 09:00 to 18:00.
Minami Aso Visitor Center: Located at Kumamoto Prefecture, Aso gun, Takamori area. Opening hours are from 09:00 to 17:00.
Climate: The Aso area can be hiked in moderate hiking shoes. However, special clothing may be needed according to diverse seasonal conditions. Winter temperature near the crater area can go down to a low of −15 °C with heavy winds and moderate snowfall, while summer is pleasant with 30–35 °C as highest during the day. June and July usually receive highest rainfall (500–600 mm).
Useful Websites:
http://www.aso-geopark.jp/ggn/pdf/annex2.pdf
http://www.aso-geopark.jp/en/about/
http://www.minamiaso-vc.go.jp (In Japanese)
http://aso-sougencenter.jp (In Japanese)

11.3.3 Challenges for Conservation of Natural Heritage

The natural heritage of the Aso area has a world-class value, but it is threatened by various anthropogenic factors. Rivers are controlled by dams, weirs, barrages, and embankments today. The planned Tateno Dam, 1 km downstream of the confluence of the Shira and Kuro Rivers, would cut a significant part of the natural input of sediments and nutrients from the watershed. The structure will have a height of 90 m, holding 10 million m^3 of water (MLIT 2007); its main aim is to reduce flood risk downstream. The dam could impact the functioning of the Shira River catchment and it would affect downstream ecosystems. Furthermore, the dam site is located in an area that has many active fault lines; this could be another major risk associated with the planned structure (Figs. 11.2 and 11.8). In the aftermath of the Kumamoto earthquake in 2016 (see more information below), major landslides took place in the area. Aso is a dynamic landscape not only for volcanism but also for changing geomorphological features and their connections to the biotic and cultural elements. The complex interaction between the different parts of this heritage should be studied in more detail and protection based on careful planning should be provided.

Timber plantations have also affected the natural and seminatural vegetation diversity of the area. Just two varieties of trees, Japanese cedar (*Cryptomeria japonica*) and Japanese cypress (*Chamaecyparis obtusa*), currently occupy a dominant part of the "forest cover." These artificial forests have very little economic value (wood price has fallen making logging unattractive), and they have "frozen" the landscape in dark, overgrown woodlands. A small stretch of primary broadleaf forest is located as a small patch in Minami Aso village known as Kitamukiyama forest and has an area of 100 ha (Kumamoto Prefectural Government 2015). Evergreen broadleaves such as *Castanopsis sieboldii, Quercus salicina Blume*, and *Quercus acuta* characterize this area.

Such recent anthropogenic changes affect the natural environment and biodiversity of Aso. Moreover, the area's status as a major tourist destination means added pressure on the natural resources. It is hoped that the concerned authorities will utilize international heritage branding schemes such as the UNESCO Geopark and the GIAHS to effectively address these challenges.

11.4 Grasslands as a Unique Biocultural Landscape

The Aso landscape is shaped by several biocultural agents. The seasonally burnt grasslands constitute one example. Approximately 22,000 ha in size, these grasslands are among the largest in Japan (FAO nd; Meyer 2017) and are regarded as landscapes of cultural significance. These landscapes are maintained by traditional knowledge and common property resource management strategies through generations. Decline of grasslands throughout Japan during the twentieth century has resulted in an effort to conserve these landscapes (MoE 2007). Coupled systems like these grasslands represent landscapes where humans and nature have coevolved for millennia, and these areas have come to be regarded as important for landscape sustainability issues (Berkes et al. 2003; Berkes 2004).

11.4.1 A Brief Ecological History of the Aso Grassland

Under natural biogeographic conditions, high rainfall and nonexistence of extensive dry periods support forested landscapes in the Aso area. During the Last Glacial Maximum (LGM), grasslands had predominated the Japanese Islands due to less humidity and precipitation. In Kyushu (which has an average precipitation of 3000 mm per year), grasslands can be observed where periodic anthropogenic disturbance such as controlled burning of vegetation took place. People living in the vicinity of the Aso caldera (Fig. 11.7) have used fire to control vegetation pattern by hindering the growth of woody trees while maintaining grasslands. Some landscapes such as marshlands were left undisturbed and natural marsh vegetation flourished in these areas.

The history of the grasslands can be determined by scientific analysis of phytolith and charcoal records in the sediments. The Aso-4 eruption widely affected land and natural vegetation; a natural vegetation succession took effect after the eruption event. Deposits of Aso-4 reveal

Fig. 11.7 Grassland with *hanaudo (Heracleum sphondylium* L. var. *nipponicum)*. The grasslands are the result of human activities coupled with geologic and geomorphic forces. Photograph by the author, July 25 2011

crystallized vegetation particles or phytoliths (SiO_2) (Miyabuchi and Sugiyama 2011). Phytoliths represent silica taken up by the plants (the silica remains even after plants decay. It is assumed that within 1000 years of Aso's last big eruption, vegetation like *Sasa* dwarf bamboo and *Zoysia* grasses came out to colonize the landscape. But during this time, the vegetation was of low density as Aso was still quite active. Gaseous eruptions obscured the sky, making it difficult for sunlight to penetrate (ibid). Miyabuchi et al. (2012) also argue that from 13,500 years BP, rapid warming climate caused some changes in the vegetation structure, but in general, *Sasa* and *Pleioblastus* dwarf bamboo dominated the landscape of the western part of Aso caldera together with sparse vegetation.

Miscanthus sinensis or pampas grass, the most dominant grass type at Aso at present, increased in the eastern side of the caldera from 13,550 BP, together with occasional stands of Quercus and Fagus trees. It is difficult for *M. Sinensis* to maintain a steady population for a long period of time although it can colonize landscape quite fast (Yamane 1973). However, Miyabuchi et al. (2012) have identified traces of continuous burning and grassland presence in the northern rim of the caldera since early Holocene and also in the eastern part of the caldera since 13,500 years ago (ibid). These traces, they argue, could be related with prehistoric grassland burning and maintenance. While it is not clear why ancient Paleolithic and Jomon people opted for such strategies, it is likely that active volcanism played a role. It is probable that foothill forests and grasslands of Aso were at first burnt due to volcanism, and then humans perpetuated this burning through controlled fire, which also benefited their livelihoods. According to studies done by the Research Institute of Humanities and Nature (RIHN), most of the human remains since the Jomon period are found from the Aso lava plateau (not the caldera area).[2] In the medieval ages, the Aso grasslands were regarded as a manor of the Aso shrine and were used for hunting rituals. In early spring, grasslands were burnt, and deer and wild boars were hunted. Grasslands were also used for grazing horses. Eventually, in modern times (after 1600 AD), grasslands became a common property and they are currently used as grazing lands for cows and horses (main usage).

The annual springtime hunting activities of the medieval times were possible due to the abundance of wild animals in the grasslands. These activities also mark the natural regeneration capacity of the land. Later, in the modern times, large-scale hunting activities were gradually replaced by a more pasture type usage with common property values. This change was associated with a considerable reduction of species variety in the grassland landscapes. A lot of grassland was converted to rice lands (a process known as *shindenkaihatsu*) during the early seventeenth century. Grasslands near the village were converted to rice lands and grasses were harvested from unused lands between the villages (Suga et al. 2012). *Shindenkaihatsu* caused deterioration of wild grasslands near villages. Moreover, in the village domain, varieties of grass were carefully selected. Careful cutting and burning avoided wild grasses that could invite insects harmful to rice and animals such as mice. Big trees were also removed from the vicinity of rice paddies to minimize shade. Thus, the grassland landscape associated with the village domain changed in modern times due to rice cultivation and its intensity.

Although the Aso grassland is among the largest in Japan, it is a threatened landscape as well (Table 11.1), and several of the uniquely adapted species are currently affected. Most of these species thrive in the pasturelands, mowing fields, and pampas grasslands (Table 11.2). For example, the perennial herb *Polemonium kiushianum* (Hanashinobu) has only about 400 remaining wild individual plants (Yokogawa et al. 2009). Habitat fragmentation and genetic hybridization with other nonnative species has had an adverse effect for the survival of this species (Matoba et al. 2011). The Large Shijimi blue butterfly (*Shijimiaeoides divinus asonis)* thrives in the mowing fields and pasturelands (Koda 2010). The Shijimi blue butterfly's life cycle depends on poisonous *Sophora flavescens* grass, which is not consumed by cattle and horses because of toxicity. The *Maculinea teleius*, another vulnerable butterfly species, thrives in the managed grasslands.

Table 11.1 Endangered species of plants of the Aso grasslands

Plant Name	Status
Hanashinobu *(Polemonium kiushianum)*	CR
Kerurisou *(Trigonotis radicans)*	EN
Tsukushitorano *(Pseudolysimachion kiusianum)*	EN
Yatsushiroshou *(Campanula glomerata var. dahurica)*	EN
Himeyuri *(Lilium concolor)*	EN
Tsukushimatsumoto *(Lychnis sieboldii var.spontanea)*	EN
Tsukushikugaisou *(Veronicastrum sibiricum var. zuccarini)*	EN
Okinagusa *(Pulsatilla cernua)*	VU
Nokaramatsu *(Thalictrum simplex var. brevipes)*	VU
Kisumire *(Viola orientalis)*	VU
Sakurasou *(Primula sieboldii)*	VU
Murasakisenburi *(Swertia pseudochinensis)*	VU
Shion *(Aster tataricus)*	VU
Michinokufukujuso *(Adonis multiflora)*	VU
Inuhagi *(Lespedeza tomentosa)*	VU
Suzusaiko *(Cynanchum paniculatum)*	VU
Kisewata *(Leonurus macranthus Maxim)*	VU
Asonokogirisou *(Achillea alpina ssp. subcartilaginea)*	VU
Hosobaoguruma *(Inula linariifolia)*	VU
Asotakarakou *(Ligularia fischeri var. takeyuki)*	VU
Hanakazura *(Aconitum ciliare)*	VU

Based on Japan Red Databook (2000) in JIBIS (nd)

[2] http://www.chikyu.ac.jp/retto/ in Suga et al. (2012).

Table 11.2 Types and characteristics of Aso grasslands

Types of grasslands	Human component	Example of characteristic species	Characteristics/uses
Mowing fields (*saisou*)	Periodic disturbances with seasonal burning	Table 11.1	Winter and summer fodder for cattle and horses
Pasture lands (*houbokuchi*)	-do-	Table 11.1	Seasonal burning, and grazing, stamping activities
Pampas grasslands (*Kaya*)	-do-	Table 11.1	Burning only, grasses for roof thatching, organic fertilizers (Washitani 2003: 21)
Marshland grasses (*Shicchisei shokubutsu*)	Natural grasslands with no disturbances	Isachne globosa, Thelypteris palustris, Drosera rotundifolia Linn., Habenaria radiata, Geranium soboliferum var. kiusianum, Aster maackii Regel, Primula sieboldii, Lychnis kiusiana Makino	Species diversity due to natural niche formation
Altered grasslands (*Kairyo-souchi*)	Altered permanently	Monoculture of Clover (Trifolium repens), and Orchard grass (Dactylis glomerata L.).	Biodiversity poor, quick fix solutions without functional connection to landscape diversity

11.5 The Ecosystem Services Connection

Currently, there are two international labels associated with the natural and managed components of Aso's heritage, namely, the UNESCO Global Geopark and the FAO Globally Important Agricultural Heritage (GIAHS). While the geopark mainly focuses on geology, the GIAHS focuses on the human management of grasslands (secondary nature). However, as this chapter argued, the geological landforms and the natural or managed features of the landscape (such as grasslands) are interrelated. Integration of these two themes is thus needed to manage a dynamic landscape like Aso where periodic human interference has shaped the ecology to a significant extent. From a landscape perspective, natural and seminatural areas are ecologically more sustainable and economically beneficial than converted ones (Balmford et al. 2002). One way to understand the importance of natural ecosystems is by looking at the different Ecosystem Services (ES) they produce (Daily 1997 In Atkins et al. 2011). Analyzing the importance of landscapes through ES has recently gained popularity as an approach in landscape conservation and management. Both material and nonmaterial benefits may be provided by ES (Chan et al. 2012). At Aso, controlled burning of the grasslands with stockbreeding and agricultural practices maintain the landscape mosaic and produce a variety of ecosystem services. Such traditional landscapes in Japan are known as "satoyama.[3]" For detailed information on these landscapes, works of Takeuchi et al. (2003), Fukamachi et al. (2001), and Yokohari and Bolthouse (2011) may be consulted; the Japan Satoyama-Satoumi Assessment is another compilation that describes the nature and status of such landscapes all over Japan (Duraiappah et al. 2012), especially with the addition of coastal and marine ecosystems (Satoumi), which are further described in Yanagi (2013).

From the ES perspective, some particular aspects of the Aso landscapes stand out. The black soil of Aso is ideal for grasslands and agricultural landscapes. Agriculture and stockbreeding create considerable economic benefit as together they generate about 29 billion JPY from the Aso landscape (FAO nd). Indirect economic benefits (which cannot easily be measured by monetary values) are derived from this landscape as well (Okubo 2013). The grasslands have major influence on the important regulating and supporting services such as the water cycle (some were explained in the geology and geomorphology section). Grassland landscapes have important habitat support functions aided by the denudation works of rivers such as Shira and Midori, which come out from the caldera. The grasslands serve as an important carbon sink; they may sequester about 32 kg of carbon per hectare per year (Toma et al. 2013). Thus, decrease of grassland ecosystems may lead to decrease in other important ecosystem functions. Studies from other parts of the world indicate that fragmentation of such ecosystems is associated with the decrease of pollinators, which in turn affect plant reproductive process. This can have negative influences for both agricultural fields and the forested landscapes (Jakobsson and Ågren 2014). Eriksson et al. (2002) observe that seminatural grasslands may house a wide variety of species, which can be as many as 50 species per meter square. Other studies find similar results from Europe (Hector et al. 1999; Loreau and Hector 2001) and the USA (Tilman and Downing 1994). This diversity is seen as vital for grassland ecosystems and their resilience (ibid).

Cultural significance of grasslands of Aso includes flower gathering during the traditional *Obon* festival (involves flowers decorations in the graves of the ancestors). Wildflowers that bloom in the grasslands after mid-August

[3] Satoyama is termed for traditional landscapes of Japan with agriculture as its mainstay. Such landscapes are usually found in close proximity of hills and low mountains with considerable forested areas.

are all regarded as *Obon* flowers. The grassland landscapes also serve as playing grounds for children. Grassland biodiversity of insects, frogs, snails, and fish in the drainage ditches can be an important recreation asset (Suga et al. 2012), with high intrinsic and educational value. The scenic beauty of rolling grasslands and a mix of active and extinct volcanoes provide an important tourist capital. According to the Kumamoto Prefecture documents, more than 17 million tourists visited the Aso landscapes in 2013 (Kumamoto Prefecture 2013). These cultural services cannot be thought separately from provisioning services as they are attached to good mental and physical health conditions of the local populace. Regulating and supporting services are also involved with sustaining and releasing of energy that is vital for all the other types of ES; a geographical perspective of ES analysis is thus an important step for ecosystem-based policy and management (Postchin and Haines-Young 2011). Such evaluation, policymaking, and implementation could be aided by international heritage labels such as the Geopark and the GIAHS. Integration of the geological, geomorphological, and biocultural components of this heritage system is required for this purpose, even as more research is required to understand the connections between them and to clarify to what extent key species and resource conditions are threatened.

11.6 The Kumamoto Earthquake and Aso Phreatic Eruption of 2016

On April 16 of 2016, the Kumamoto Prefecture was hit by a magnitude 7 earthquake that damaged a large part of the Kumamoto city and its vicinities. The tremor was a part of a series of quakes that shook Kumamoto for several days. About 90,000 people became displaced from their home and 49 people lost their lives. Crustal movements along two fault lines (Fig. 11.8), the Futagawa and

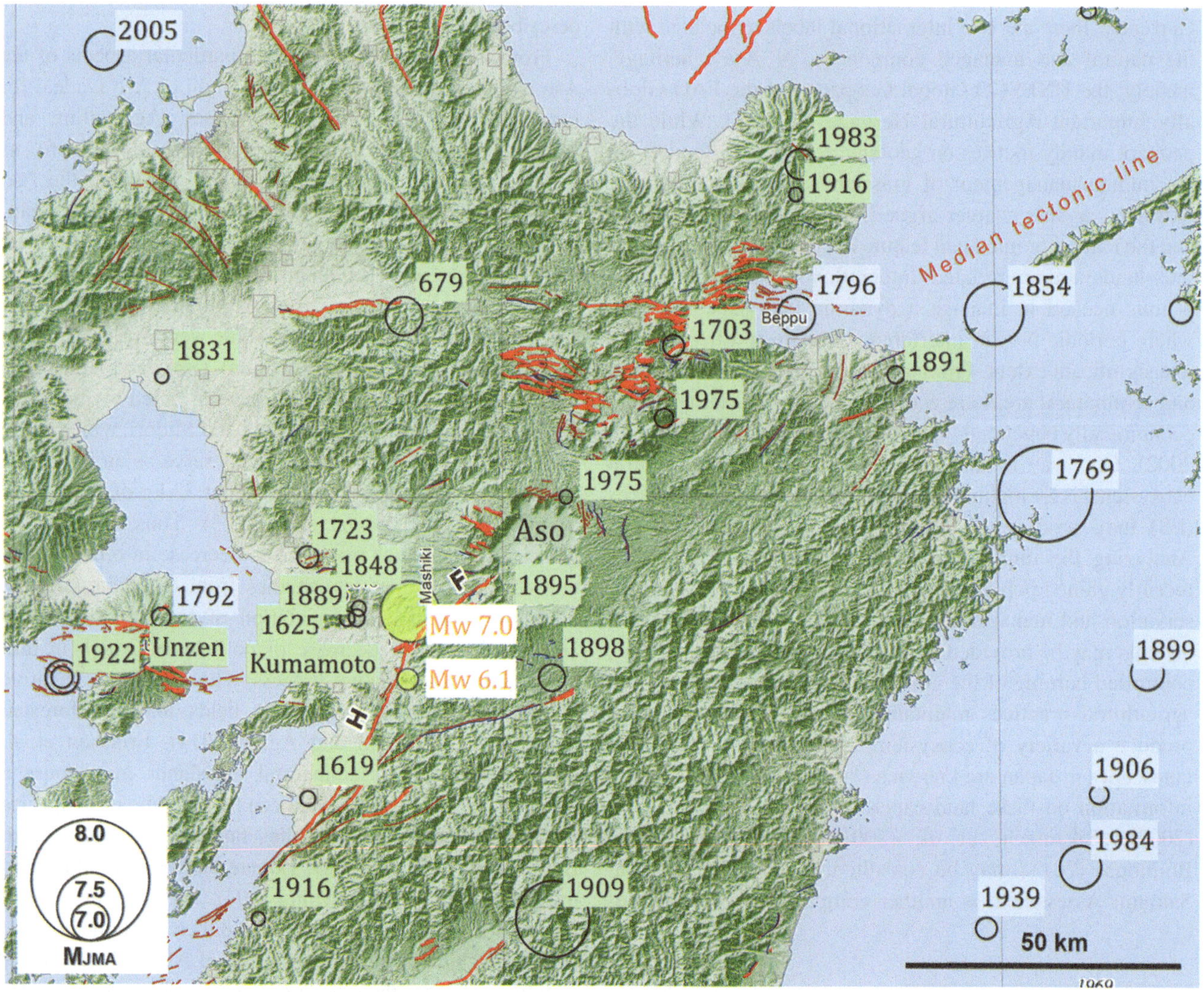

Fig. 11.8 Image of fault systems in Kyushu. Source: Okumura (2016). Reproduced with permission

Hinagu, along the Central Kyushu Rift, caused these tremors (Okumura 2016). The Aso area sits in the middle of this rift graben between the Unzen volcano area and the Beppu Haneyama area. The Aso Volcano Museum and the roadways that connect to the museum were damaged, and the access to Aso geopark was temporarily blocked.

On 8 October 2016, a large phreatic eruption took place at the Mt. Naka crater of Aso. The resulting ash column rose 11 km into the air (CNN 2016), and the Aso area was affected by moderate to severe ashfall. At present, there is a continued possibility that more large eruptions could follow (ibid). However, the eruption was a phreatic event and no upwelling of magma was detected.

Although such natural hazards are related to human sorrow and economic loss, these processes are vital for landscape formation and evolution. Damage to life and property can possibly be reduced by making more landscape-conscious development pathways. This is especially true for places such as Aso, where crustal movements and volcanic landscapes create a geomorphologically unstable but ecologically rich heritage.

11.7 Conclusion

Geology, geomorphology, ecology, and culture are all parts of the natural heritage at Aso. These social-ecological landscapes, with their broader geological and geomorphological attributes, help us understand landscapes and their conservation in a more holistic manner. In the case of Aso, one significant biocultural component, the grasslands, clearly shows a marked decline during modern times. This chapter provided an outline of the geological heritage of Aso and also analyzed how a seminatural component of the complex heritage can provide vital clues for understanding connections between geology, geomorphology, ecology, and culture at multiple levels. Knowledge acquired from such landscapes can be used for increasing awareness for more holistic planning and conservation. For holistic conservation, the need is to create awareness about the connectivity and functions of landscapes through considering them as a part of the broader geological and geomorphological landform mosaic. An ES framework may help in both creating this awareness as and providing insights into the linkages between abiotic and biotic factors. This is especially necessary in order to achieve the common goal of fully appreciating and understanding the landscape and landform diversity that sustains life over many generations.

References

Aso UNESCO Global Geopark (2012a) Highlights of Aso Geopark. http://www.aso-geopark.jp/en/charm/index.html. Accessed 23 Apr 2017

Aso UNESCO Global Geopark (2012b) Annex 2. Geological Heritage. Aso Geopark Promotion Council, Japan. http://www.aso-geopark.jp/ggn/pdf/annex2.pdf. Accessed 16 Aug 2016

Aso UNESCO Global Geopark (2012c) Geosite introduction of Aso: Furubochu Geosite. http://www.aso-geopark.jp/en/geosites/geosite05.html. Accessed 20 Mar 2017

Aso UNESCO Global Geopark (2012d) Geosite introduction of Aso. http://www.aso-geopark.jp/en/geosites/geosite05.html

Atkins JP, Burdon D, Elliott M, Gregory AJ (2011) Management of the marine environment: integrating ecosystem services and societal benefits with the DPSIR framework in a systems approach. Mar Pollut Bull 62:215–226

Balmford A, Bruner A et al (2002) Economic reasons for conserving wild nature. Science 297:950–953

Berkes F (2004) Rethinking community-based conservation. Conserv Biol 18:621–630

Berkes F, Colding FJ, Folke C (2003) Navigating social-ecological systems: building resilience for complexity and change. Cambridge University Press, Cambridge

Chan K, Guerry A, Satterfield T et al (2012) Integrating 'cultural' and 'social' into ecosystem services: a framework for value-based ecosystem decision-making. BioScience 62:744–756

Chapman N, Apted M et al (2009) Development of methodologies for identification of volcanic and tectonic hazards to potential HLW repository sites in Japan: the Kyushu case study. Nuclear Waste Management Organization of Japan (NUMO). http://www.numo.or.jp/en/reports/pdf/TR-09-02.pdf. Accessed 11 Nov 2016

CNN (Cable News Network) (2016) Japan: Mount Aso volcano belches out 11,000-meter ash cloud. http://edition.cnn.com/2016/10/08/asia/japan-volcano-eruption-mount-aso/. Accessed 11 Nov 2016

Daily GC (1997) Nature's services: societal dependence on natural ecosystems. Island Press, Washington, DC

Duraiappah AK, Nakamura K, Takeuchi K, Watanabe M, Nishi M (eds) (2012) Satoyama-satoumi ecosystems and human well-being: socio-ecological production landscapes of Japan. United Nations University Press, Tokyo

Eriksson O, Cousins SO, Bruun HH (2002) Land-use history and fragmentation of traditionally managed grasslands in Scandinavia. J Veg Sci 13:743–748

Erikstad L (2013) Geoheritage and geodiversity management: the questions for tomorrow. Proc Geol Assoc 124:713–719

Food and Agricultural Organization (FAO) (nd) Managing Aso grasslands for sustainable agriculture. http://www.fao.org/3/a-bp797e.pdf. Accessed 23 Nov 2016

Food and Agricultural Organization (FAO) (2017) Globally Important Agricultural Heritage Systems (GIAHS): Asia and the Pacific. http://http://www.fao.org/giahs/giahsaroundtheworld/designated-sites/asia-and-the-pacific/en/. Accessed 23 Apr 2017

Fukamachi K, Oku H, Nakashizuka T (2001) The change of satoyama landscape and its causality in Kamiseya, Kyoto Prefecture, Japan. Landsc Ecol 16:703–717

Geological Survey of Japan, National Institute of Advanced Industrial Science and Technology (GSJ, AIST) (2017) Kuju volcano. Active volcanoes of Japan. https://gbank.gsj.jp/volcano/Act_Vol/kuju/text/eng/exp-3e.html#. Accessed 4 Feb 2017

Gray M (2013) Geodiversity: valuing and conserving abiotic nature. Wiley, Somerset

Hector A (1999) Plant diversity and productivity experiments in European Grasslands. Science 286(5442):1123–1127

Jakobsson A, Ågren J (2014) Distance to semi-natural grassland influences seed production of insect-pollinated herbs. Oecologia 175:199–208. doi:10.1007/s00442-014-2904-z

Japan Seismic Hazard Information Station (JSHIS) (2011) http://www.j-shis.bosai.go.jp/map/. Accessed 23 Nov 2016

JIBIS (Japan Integrated Biodiversity Information System) (nd) Red list of threatened plants of Japan. http://www.biodic.go.jp/english/rdb/rdb_f.html . Accessed 2 Dec 2016

Kamata H, Kobayashi T (1997) The eruptive rate and history of Kuju volcano in Japan during the past 15,000 years. J Volcanol Geomorphol Res 76:163–171

Koda K (2010) Analysis of the geographical distribution of an endangered butterfly, the Large Shijimi Blue, Shijimiaeoides divines, using insect database. J Fac Agric Shinshu Univ 46(1–2):99–103

Kumamoto City Website (2012) Kumamoto shi no Jinkou, Setaisu Ichirann Hyou (Population and housing statistics of Kumamoto City). http://www.city.kumamoto.kumamoto.jp/tokei/content/ASP/Jinkou/default.asp. Accessed 2 Dec 2016 (In Japanese)

Kumamoto Prefectural Government (2015) Kitamukiyama no genshirin: minamiaso mura (The primary forests of Kitamukiyama, Minamiaso village). http://www.pref.kumamoto.jp/kiji_8140.html. Accessed 10 Nov 2016 (In Japanese)

Kumamoto Prefecture (2013) Kumamoto ken kankou toukei hyo. (Tourism statistics of Kumamoto Prefecture). https://www.pref.kumamoto.jp/common/UploadFileOutput.ashx?c_id=3&id=6591&sub_id=1&flid=1&dan_id=1 Accessed 2 Dec 2016 (In Japanese)

Loreau M, Hector A (2001) Nature 412(6842):72–76

Machida H, Arai F (1983) Extensive ash falls in and around the sea of Japan from large late quaternary eruptions. J Volcanol Geotherm Res 18(1):151–164

Matoba H, Inaba K, Nagano K, Uchiyama H (2011) Use of RAPD analysis to assess the threat of interspecific hybridization to the critically endangered *Polemonium kiushianum* in Japan. J Plant Res 124:125–130. doi:10.1007/s10265-010-0353-2

Meyer MH (2017) Japan's miscanthus grasslands. Miscanthus. College of Foof Agricultural and Natural Resources Sciences, University of Minnesota. http://miscanthus.cfans.umn.edu/habitat/miscanthus-japan. Accessed 2 Feb 2017

Miyabuchi Y (2009) A 90,000-year tephrostratigraphic framework of Aso volcano, Japan. Sediment Geol 220:169–189

Miyabuchi Y, Sugiyama S (2011) 90,000-year Phytolith record from tephra section at the northeastern rim of Aso caldera, Japan. Quat Int 254:73–82

Miyabuchi Y, Sugiyama S, Nagaoka Y (2012) Vegetation and fire history during the last 30,000 years based on phytolith and macroscopic charcoal records in the eastern and western areas of Aso Volcano, Japan. Quat Int 254:28–35

Miyoshi M, Shinmura T, Sumino H et al (2013) Lateral magma intrusion from a caldera-forming magma chamber: Constraints from geochronology and geochemistry of volcanic products from lateral cones around the Aso caldera, SW Japan. Chem Geol 352:202–210. doi:10.1016/j.chemgeo.2013.06.003

MLIT (Ministry of Land, Infrastructure, Transport and Tourism) (2007) Tateno dam to wa? (Tateno dam). http://www.qsr.mlit.go.jp/tateno/damujigyo/aramasi03.html. Accessed 10 Nov 2016 (In Japanese)

MOE (Ministry of the Environment) (2007) Aso sougen saisei zentai koso: Aso sougen wo mirai e (Aso grasslands for the future: comprehensive plan for Aso grassland restoation). https://www.env.go.jp/nature/saisei/law-saisei/aso/aso0_1.pdf. Accessed 20 Mar 2017

Nakai Y, Shindo T, Fukuoka T, et al (nd) Aso chiiki iseki no kokuyou seki sekki no gen seki sanchi. http://ci.nii.ac.jp/els/110006213649.pdf?id=ART0008234741&type=pdf&lang=en&host=cinii&order_no=&ppv_type=0&lang_sw=&no=1486364814&cp

NASA Shuttle Radar Topography Mission (nd) https://commons.wikimedia.org/wiki/File:Aso_Caldera_3D_2012.jpg. Accessed 10 Nov 2016

Okubo K (2013) An economic valuation of the grassland landscapes in Aso Kuju national park, Japan. In: Martijn van der Heide C, Heijman W (eds) Economic value of landscapes. Routledge, London, pp 189–203

Okumura K (2016) Earthquake geology of the April 14 and 16, 2016 Kumamoto earthquakes. In Epstein W, Nishimoto T, Okumura K, Swan S, Yanev P (eds) Kumamoto earthquake investigation: a preliminary report. http://janet-dr.com/11_saigaiji/The%20Kumamoto%20Earthquake%20Investigation_%20A%20Preliminary%20Report.pdf. Accessed 11 Nov 2016

Ono K, Watanabe K (1985) Geological map of Aso volcano (1:50,000). Geological Map of Volcanoes 4, Geological Survey of Japan. Geological Survey of Japan, AIST. https://gbank.gsj.jp/volcano/Act_Vol/aso/text/eng/exp04-1e.html. Accessed 2 Dec 2016 (In Japanese with English abstract)

Postchin MB, Haines-Young RH (2011) Ecosystem services: exploring a geographical perspective. Prog Phys Geogr 35(5):575–594

Suga T, Okamoto T, Ushimaru A (2012) Souchi to nihonjin: nihon retto sougen 1 man nen no tabi (Grassland and Japanese: a journey of 10,000 years of Japan's grasslands). Chikuchi shokan: Tokyo (In Japanese)

Takahashi Y (2014) Kusa no Satoyama to Ikiru (Living with Grassland Satoyama). Ecological Society of Japan. https://www.esj.ne.jp/esj/book/ecology_7/05_takahashi.pdf. Accessed 2 Dec 2016 (In Japanese)

Takeuchi K, Brown RD, Washitani I, Tsunekawa A, Yokohari M (eds) (2003) Satoyama: the traditional rural landscape of Japan. Springer, Tokyo

Tilman D, Downing A (1994) Biodiversity and stability in grasslands. Nature 367:363–365

Toma Y, Brown JC, Sugiyama S, Nakaboh M, Hatano R, Fernández FG, Stewart RJ, Nishiwaki A, Yamada T (2013) Soil carbon stocks and carbon sequestration rates in seminatural grassland in Aso region, Kumamoto, Southern Japan. Glob Chang Biol 19(6):1676–1687

Washitani I (2003) Satoyama landscapes and conservation ecology. In: Takeuchi K, Brown RD, Washitani I, Tsunekawa A, Yokohari M (eds) Satoyama: the traditional rural landscape of Japan. Springer, Tokyo, pp 16–23

Watanabe K, Ono K (1969) Geology of the vicinity of Omine on the western flank of the Aso caldera. J Geol Soc Jpn 75:365–374 (In Japanese with English abstract)

Wild Bird Society of Japan (2009) Shirakawa Estuary. http://www.wbsj.org/nature/hogo/others/iba/eng/144.html. Accessed 2 Dec 2016

Yamane I (1973) The meaning of Miscanthus sinensis in the formation of Kuroboku soil. Pedologist 17:84–94 (In Japanese with English summary)

Yanagi T (2013) Japanese commons in the coastal seas: How the satoumi concept harmonizes human activity in coastal seas with high productivity and diversity. Springer, Japan

Yokogawa M, Kaneko S, Isagi Y (2009) Development of microsatellite markers for Polemonium kiushianum (Polemoniaceae), a critically endangered grassland plant species in Japan. Conserv Genet 10:1445. doi:10.1007/s10592-008-9757-0

Yokohari M, Bolthouse J (2011) Keep it alive, don't freeze it: a conceptual perspective on the conservation of continuously evolving satoyama landscapes. Landsc Ecol Eng 7(2):207–216

The Mt. Unzen Disaster: A Terrible Learning Experience

12

Malcolm J.M. Cooper

Abstract

Mount Unzen is an active composite stratovolcano in the Southwest Japan Arc. It is made up of a cluster of overlapping volcanoes that have formed at various times over the past 6 million years on the Shimabara Peninsula of Kyushu Island. This complex is close to the major cities of Nagasaki and Kumamoto and has considerable settlement around it on the peninsula. Of course, communities of human beings have always lived on volcanoes and their surrounding landscapes, whether active or dormant. These volcanoes provide visitors to such communities with access to their geological history and have even provided the remains of previous settlements now covered by eruptions (e.g., Pompeii (Vesuvius AD79) and Soufrière Hills (Montserrat). As a representative of the most dangerous of these volcanoes, Mt. Unzen became a worldwide media sensation in 1991 when it produced a series of massive eruptions (which claimed 44 lives) after many years of calm. This event raised concerns about community awareness of volcanic activity and created new opportunities for the study of major volcanic events and their precursors. From this situation, some of the more destructive volcanoes have been classified as requiring constant monitoring (the *Decade Volcanoes*), while calls for more information on all those identified as active (and, more recently, on those considered dormant given several occurrences of reactivation that were not considered likely) are frequently made.

Keywords

Mt. Unzen • Decade volcanoes • Eruption cycles and impacts • Geology

12.1 Introduction

The *Cities on Volcanoes 5th annual conference* (COV5) first alerted this author to the extensive role and impact that natural heritage has in the communities of Japan. This conference was held in the town of Shimabara, Kyushu, situated on the flanks of *Unzendake* (Mt. Unzen), in the Ariake Sea, and close to the cities of Kumamoto and Nagasaki. Mt. Unzen and its companion in historical and potential future destructiveness on the Japanese island of Kyushu, Mt. Sakurajima (near Kagoshima), were classed as "Decade Volcanoes" in the 1990s (Siebert and Simkin 2002) and thus became part of a monitoring project run by the International Association of Volcanology and Chemistry of the Earth's Interior (IAVCEI) during the International Decade for Natural Disaster Reduction (The 1990s—United Nations Resolution 44/236 1989). This project led to a subcommission of the IAVCEI being appointed to:

> direct attention to a small number of dangerous active volcanoes world-wide, and to encourage the establishment of a range of research and public-awareness activities aimed at enhancing an understanding of the volcanoes and the hazards posed by them (Siebert and Simkin 2002).

In Japan, over 100 volcanoes are currently classed as active (Kuno 1962; Erfurt-Cooper and Cooper 2010). The Japan Meteorological Agency (JMA) defines volcanoes that

M.J.M. Cooper (✉)
Ritsumeikan Asia Pacific University, Beppu, Japan
e-mail: cooperm@apu.ac.jp

A. Chakraborty et al. (eds.), *Natural Heritage of Japan*, Geoheritage, Geoparks and Geotourism,
DOI 10.1007/978-3-319-61896-8_12

have been active at some time during the past 10,000 years, or which have active fumaroles, as active volcanoes. To indicate the degree of volcanic activity, the JMA devised an alert system in November 2003 (0–5; see Table 12.1) for active volcanoes, many of which are continuously monitored. When necessary, information on volcanic activity is given to disaster prevention authorities and to the public so that they might undertake relevant disaster reduction measures (JMA 2007). This is especially so with the island of Kyushu. This island is the southernmost of the four major islands of Japan, and like Honshu and Hokkaido, it is home to several active volcanoes (at least 7; Table 12.2), many of which are constantly erupting. Many of these volcanically active locations double up as major tourist attractions, and in cases where they are located within geoparks (both UNESCO Geoparks and Japan's national geoparks), they constitute the defining geosites of those areas. All these geosites are tourist destinations, all are also located close to urban areas, and all have a recent eruption history as outlined in Table 12.2.

Table 12.1 Volcanic Activity Level classification by the Japan Meteorological Agency

Level	General activity (Detailed activity is listed by JMA for each volcano as required)
5	*Very large eruption* Danger over wide areas
4	*Large eruption* Danger also in specified areas outside the vent and its surrounds
3	*Eruption* Minor eruption or a precursor to large eruptions, danger areas nearby or confined to the crater
2	*Active Volcanism* Seismic swarm, tremors, or slight eruptions, considered preliminary to an eruption
1	*Calm* Minor seismic activity, possibly active fumaroles, but no signs of an impending eruption
0	*Dormant* No active fumaroles, no seismic activity for a long period

Source: JMA (2007)

Table 12.2 Active volcanic areas in Kyushu close to large cities

Volcano	Elevation (meters)	Last erupted	Nearby cities (population)
Mount Aso	529	Ongoing	Kumamoto (670,000)
Mount Kuju	1788	1996	Beppu and Oita (124,000/475,000) Kumamoto
Tsurumi Lava Dome Complex	1584	867	Beppu and Oita
Unzen stratovolcano complex	1500	1991–1996	Shimabara (50,000), (Nagasaki 450,000) Kumamoto
Sakurajima	1117	Ongoing	Kagoshima (605,000)
Kirishima	1700	2011	Kagoshima
Kaimondake	922	885	Kagoshima

Source: The author

We note throughout this book that natural heritage is an integral part of the human landscape. However, in using these resources, the impact of many of the processes involved in their creation and ongoing development have been catastrophic for mankind, for example, events such as earthquakes, floods, or volcanic eruptions. The understanding of these phenomena requires geoscience, but dealing with them also involves communities. In the past few years, scientific issues like volcanic activity, earthquakes, and floods in settlement areas have engaged the attention of political decision-makers, for the very simple reason that the community impacts of many of them are impossible to disguise. But more fundamentally, we all need to understand the importance of natural heritage, and the increasing importance of conserving it in all countries, if we are to provide for our own and future generations' natural capital.

The growth of visitor interest in these natural heritage resources has led to the redefinition of sector labels like ecotourism and world heritage lists to incorporate geology and volcanology into tourism, and the coining of the new title, *Geotourism* (Dowling and Newsome 2010; Erfurt-Cooper and Cooper 2010). These new approaches are slowly being integrated into the marketing of individual regions; for example, the *Shimabara Volcanic Area*, which includes Mt. Unzen, was officially recognized as a Global *Geopark* in August 2009 (McKeever 2009). The most important features for this geopark are its volcanic and geothermal attractions, as well as several volcano museums, a village buried by lahars, and the natural environment linked to these features. Several regions in Japan are also existing or aspiring UNESCO Global Geoparks, or National Geoparks (Japanese Geoparks Network), based on volcanic and geothermal landforms. Volcanic features are thus among the greater draw cards in Japanese national and prefectural parks.

12.2 Living with Volcanoes

12.2.1 Volcanoes as Geological Natural Heritage

Communities of human beings have always been attracted by volcanoes and their surrounding landscapes (Erfurt-Cooper 2014), whether active or dormant. Volcanoes have provided visitors to such communities with experiences covering their history and even previous settlements covered by eruptions [e.g., Pompeii (from Vesuvius in AD79), Soufrière Hills (Montserrat), Mt. St Helens (USA, 1980s), and others], or their cultures, often as depicted by artists (e.g., the almost perfect cone shape of Mount Fuji). Some are accessible as tourist attractions, but others are in more remote locations (e.g., in Alaska, Antarctica, and the Kamchatka Peninsula of Russia). Some are destinations because of specific features (e.g., Erta Ale lava lake in Ethiopia), while others are simply in the visited landscape (Erfurt-Cooper 2014). All these sites have associated with them materials that illustrate the geological natural heritage associated with an area, and this makes them ideal as educational tools as well as illustrations of the forces at work.

We can further illustrate these influences using any one of a great number of individual volcanoes and their associated communities, but it is important for the development of our sense of the influence that volcanoes can have on the development of natural capital and heritage that we cover the most dangerous of these and their impact. As a representative of these volcanoes, Mt. Unzen became a worldwide media sensation in 1991 when it produced a series of massive eruptions (which claimed 44 lives) after many years of calm. This event raised concerns about awareness of volcanic activity and created new opportunities for the study of major volcanic events and their precursors (Nakada et al. 1999; Hoshizumi and Nakano 2009; Erfurt-Cooper 2014). From this situation, some of the more destructive have been classified as requiring constant monitoring (the *Decade Volcanoes*), while from time to time there are calls for more information on all those identified as active (and, more recently, on those considered dormant given several occurrences of reactivation that were not considered likely).

Decade volcanoes are a set of 16 identified by the IAVCEI as requiring constant attention in the "light of their history of large, destructive eruptions, and their proximity to populated areas" (Erfurt-Cooper 2014). The IAVCEI encourages research and awareness programs at these sites, with the twin aims of reaching a better understanding of the problems they present, and being able to reduce the severity of any disaster that may occur as an outcome of their continued activity (Cooper 2014). A site may be listed as a Decade Volcano if it presents more than one major hazard over time (ashfalls, pyroclastic and lava flows, lahars, and lava dome collapses—all of which the Unzen Volcano experienced between 1991 and 1996 and many times in the past; shows recent geological activity; is close to a populated area (eruptions may threaten large populations, and therefore lessening eruption hazards at these geosites is critical); is politically and physically accessible for continued study; and there is local support for the required research and policy making.

Table 12.3 lists a selection of the more recent major eruptions around the world (Smithsonian Institute 2015). The database that these were drawn from contains over 1430 confirmed eruptions in Japan since the sixteenth Century and 9726 events worldwide since 10,000 BCE (Erfurt-Cooper 2014)—this is natural heritage at its rawest in respect of human communities, but it is important to base policy on the fact that in the twenty-first century *more* activity will boost attractiveness to tourists instead of reducing the volcano's popularity, as compared to "normal" activity or dormancy (Tilling 1989; King and Brattstrom 2014), leading to questions about appropriate risk management for tourists of this form of heritage.

12.2.2 The Geological Basis of Volcanism in Japan

Japan is an archipelago of many islands that has a geological history of over 500 million years and is located at the

Table 12.3 Examples of very recent volcanic eruptions

Some recent major eruptive events (location, date, and type) since 2011	
2008–11 Chaitén, Chile (pyroclastic flows, lava dome)	2012 Stromboli, Italy (ash eruption)
2011 Nyiragongo, DR Congo (lava lake)	2012 Karymsky, Kamchatka, Russia (lava, ash)
2011 Eyjafjallajökull, Iceland (fissure, ash eruption)	2012 White Island, New Zealand (crater lake, ash)
2011 Grímsvötn, Iceland (subglacial)	2012 Tongariro, New Zealand (ash, pyroclasts)
2011 Erebus, Antarctica (lava lake)	2012–Yasur, Vanuatu (strombolian, ash)
2011 Bromo, Indonesia (ash, pyroclasts)	2013 Galeras, Colombia (ash, gas)
2011 Ulawun, Papua New Guinea (gas, steam)	2011–14 Merapi, Indonesia (lava dome, pyroclastic flows)
2011 Kilauea, Hawaii USA (lava lake)	2014 Krakatau, Indonesia (lava, ash)
2012 Soufriere Hills, Montserrat Caribbean (ash, lava dome, pyroclastic flows)	2014 Mayon, Philippines (ash, gas, lava)
2012 Nevado del Ruiz, Colombia (ash, lahars)	2014 Piton de la Fournaise, Reunion (lava flows)
2012 Puyehue Volcano, Chile (ash eruption)	2014 Rincón de la Vieja, Costa Rica (crater lake)
2012 Nyamuragira, DR Congo (lava flows, gas)	2014–15 Aso, Japan (ongoing ash eruptions)
2012 Reventador, Ecuador (ash, pyroclastic flows)	2012–2015 Lokon, Sulawesi Indonesia (ash eruption)
2012 Tungurahua, Ecuador (lava, gas, ash)	2012–15 Sakurajima, Japan (ongoing ash)
2012 Santa María/Santiaguito, Guatemala (ash, lava)	2015 Tungurahua, Ecuador (lava, gas, ash)
2012 Krakatau, Indonesia (lava dome growth)	2015 Cotopaxi, Ecuador
2012 Etna, Italy (lava, ash, strombolian)	

Source: Public domain published data from the Global Volcanism Program, Natural Museum of Natural History, Smithsonian Institution. http://volcano.si.edu/search_eruption_results.cfm#, modified by the author junction of four converging tectonic plates: the Philippine Sea, Eurasian, Pacific, and North American Plates (Machida 2015). The main islands of Kyushu, Shikoku, and Honshu (partly) occupy the leading edge of the Eurasian plate next to the Nankai Trough that marks the subduction zone of the Philippine Sea Plate, which dips westward below southern Japan. The West Japan Volcanic Zone (WJVZ) developed from this subduction, primarily in Kyushu, as well as in southwestern Honshu and smaller volcanic islands to the south. Further complications include a spreading ridge system that extends from southern Kyushu southward to Okinawa and Taiwan. This spreading zone is marked in southern Kyushu by the Kagoshima graben and the Sakurajima volcano.

This geological situation has created five volcanic arcs (Erfurt-Cooper and Cooper 2010). These arcs meet at a triple junction on the island of Honshu. The Northeast Honshu Arc and the Kurile Arc trend to the northeast. The Izu-Bonin Arc trends to the southeast. The Southwest Honshu Arc and the Ryukyu Arc trend to the south west. The Northeast Honshu Arc and the Kurile Arc have been formed by the subduction of the Pacific Plate under the North American Plate. The Izu-Bonin Arc is the result of subduction of the Pacific Plate under the Philippine Sea Plate. The Southwest Honshu Arc and the Ryukyu Arc have been formed by the subduction of the Philippine Sea Plate under the Eurasian Plate (Aramaki and Ui 1982). Japan has a long record of documented historic and inferred prehistoric eruptions resulting from this situation. Tephrochronology (measurement of elapsed time based on the nature of volcanic sediments) is commonly used for the study of arc volcanism in Japan, and this approach allows for a more detailed and balanced view of volcanic activity over the last 10,000 years. Japan leads the world's volcanic regions with 1274 dated eruptions and has 94 volcanoes with dated eruptions. Japan also leads the world with 41 large (VEI greater than or equal to 4) explosive eruptions in the last 10,000 years. Pyroclastic flows, one of the deadliest volcanic hazards, have occurred at 28% of Japan's eruptions (Seibert and Simkin 2002).

12.3 The Geology of the Mt. Unzen Complex and the 1991 Eruption

12.3.1 Overview

Mount Unzen is an active composite stratovolcano in the Southwest Japan Arc. It is in fact a cluster of overlapping volcanoes which formed at various times over the past 6 million years in the Shimabara Peninsula. There have been many disastrous explosive eruptions from this volcano, and it is currently active (Uto et al. 2001). Its most recent eruption began in 1991 and lasted until 1995. Unzen-dake got its name from all the hot springs found in its vicinity. Originally, the mountain is said to have been called "Onsen-dake," onsen means hot springs, and "dake" means peak, so the translation of Unzen-dake into English is "hot spring mountain" (Hoshizumi et al. 1999; Hoshizumi and Nakano 2009).

The area known as Mount Unzen makes up a considerable part of the Shimabara Peninsula of Kyushu (see Fig. 12.1) and is close to the cities of Nagasaki and Kumamoto. The oldest volcanic evidence is dated at 6 Ma, with further massive deposits across the whole peninsula being made between 0.5 Ma and 2.5 Ma. During the past half million years, the volcano has erupted lavas and pyroclastic materials of andesite to dacite composition and has developed a volcano-tectonic graben (Hoshizumi et al. 1999; Nakada et al. 1999; Uto et al. 2001). This evolution means that, in geological terms, the Unzen complex can be divided into Older and Younger Unzen volcanoes. The exposed rocks of the Older Unzen volcano are composed of thick lava flows and pyroclastic deposits dated around 2–3 million years. Drill cores recovered from the basal part of the Older Unzen volcano are dated at 4–5 Ma. The volume of the ejecta from the Older Unzen volcanoes exceeded 120 km^3 (Hoshizumi and Nakano 2009).

The Younger Unzen volcano is composed of lava domes and pyroclastic deposits, mostly younger than 1 million years (Hoshizumi et al. 2004). This group is made up of the existing Nodake, Myokendake, Fugendake, and Mayuyama vents. Nodake, Myokendake, and Fugendake volcanoes are 1 million to 700,000, 300,000–200,000, and <200,000 years old, respectively. Mayuyama volcano is the most recent, having been formed from lava domes on the eastern flank of the Unzen composite volcano about 4000 years before present. Total eruptive volume of the Younger Unzen volcano sequence was about 8 km^3, and the eruptive production rate is calculated as being of one order of magnitude smaller than that of the Older Unzen volcano (Uto et al. 2001).

More recent activity (0 to 150,000 years) is noted at the following sites in the volcano complex: activity at Nodake (70–150,000 years), Myokendake (25–40,000), Fugendake (less than 25,000 years ago, and now), and Mayuyama (4000 years) are particularly well known (Uto et al. 2001). The Fugendake outlet lies about 6 km from the largest settlement on the peninsula (Shimabara), and the biggest eruption before 1991 from this geosite was recorded in 1792 (a large lava flow; Erfurt-Cooper 2014). The Mayuyama dome is noted for the fact that it collapsed unexpectedly during this eruption (on 21 May), creating a landslide and causing a mega-tsunami that reached a height of 20–57 m (15,000 people are said to have died) in the adjacent Ariake Sea. At time of writing, this is the worst volcano-related death toll that Japan has experienced (Johnston 2011).

Fig. 12.1 Mt. Unzen. Source: http//academic.emporia.edu/aberjame/tectonic/japan/japan_map1.jpg. Note: Fugendake is the tallest younger volcano in the Unzen Complex and is toward the back of the photo. The city of Shimabara is in the foreground and the older volcanoes at the back. The main debris flows from Fugendake can be seen to the *left* and *right* of the massif of Mayuyama in the center

After this event, the volcano complex remained quiet until the 1990s (Nakada et al. 1999). However, it became obvious in 1990 that further eruptions might occur soon, and this threat prompted local authorities to evacuate some 12,000 people from around the volcano in 1991. On June 3, 1991, the volcano erupted violently (Erfurt-Cooper 2014). A massive and immediate 4.5 km pyroclastic flow killed 43 people including scientists and journalists, while thousands of smaller flows occurred over the period 1991 to 1995. From 1993, the eruptions gradually declined and had ceased by 1995. But since then heavy rain has frequently remobilized pyroclastic material, generating lahars, which forces local authorities to construct huge diversion structures in several of the radial river valleys on the volcano to channel these flows away from vulnerable areas. Local early-warning systems and evacuation plans have also been created.

12.3.2 The 1990–1995 Eruption Sequence

This section describes the 1990–1995 eruption sequence. Following about 200 years of no activity, a small phreatic eruption started at the summit of Unzen Volcano (Fugendake) in November 1990 (Nakada et al. 1999). A swarm of earthquakes had been felt in the area below the western flank of the volcano a year before this eruption, and isolated tremors occurred below the summit shortly before it. Following this period of initial activity, phreatomagmatic eruptions began in February 1991, becoming larger over time, and developed into a dacite dome eruption in May 1991 that lasted approximately 4 years (Nakada et al. 1999). The lava dome grew in an unstable form on the shoulder of Fugendake, with repeating partial collapses, but grew eventually to cover the crater. Magma was discharged nearly continuously through the period of its growth, and pyroclastic flows frequently descended on the northeast, east, and southeast flanks of the volcano. Major pyroclastic flows took place when the lava effusion rate was high. The total volume of magma erupted was 2.1×10^8 m^3, but about a half of this volume remained as a lava dome at the summit (1.2 km long, 0.8 km wide, and 230–540 m high). In February 1995, the eruption finished with extrusion of a spine at the endogenous dome top. The dome subsequently started slow deformation and cooling after the halt of magma effusion (Hoshizumi et al. 1999; Uto et al. 2001).

The 1990–1995 eruption was characterized by repeated lava dome growth and block and ash flows triggered by the dome collapse (Hoshizumi et al. 1999; Uto et al. 2001). The repeated dome collapse events also produced ash cloud surges and fine-grained ashfall deposits. The block and ash flow deposits, which are confined to valleys, are massive to reversely graded, are very poorly sorted, and contain non-vesiculated to poorly vesiculated blocks up to 10 m across (Miyabuchi 1999). The deposits often display finer-grained basal parts. Ash cloud surge deposits, which spread more widely, are finer grained and better sorted than the block and ash flow deposits. The ashfall deposits, derived from ash clouds, are dispersed around the volcano and are very fine grained and sorted. Lahars resulting from remobilization of this pyroclastic debris after rain devastated the base of the volcano and resulted in massive engineering works to divert them to the sea.

By the end of the eruption, a semipermanent lava dome with dimensions of 1.2 × 0.8 km and a thickness of 230–540 m had been created (Hoshizumi et al. 1999; Uto et al. 2001). Its volume approaches 0.1 km^3, and a slightly greater amount of deposits attributable to the eruption (pyroclastic and debris flow deposits) is spread around its flanks. In total, about 0.21 km^3 of magma was erupted from Fugendake (Miyabuchi 1999). The lava dome consists of plagioclase-phyric dacite with silicate content of between 64.5% and 66% and was erupted at temperatures of 780–870C. Effusion rates of about 7 m^3 per second were recorded in 1991, with a second minor peak of about 2.5 m^3 in 1993. Sulfur dioxide emissions also peaked near the beginning of the eruption at a level of about 250tonnes/day. The summit of the lava dome resulting from the eruption is at 1483 m and is now referred to as Heisei Shinzan (which dates it for posterity in the current Emperor's reign). It is to the east of the former summit of Fugendake (1359 m).

Nearly 10,000 Pyroclastic Flow (PF) events were detected during the eruption (Yamamoto et al. 1993; Fig. 12.2). All but one of these were block and ash flows resulting from collapse events at the lava dome. However, on 8 June 1991 a vulcanian type explosion resulted in a pumiceous pyroclastic flow which was particularly mobile, surpassing all previous flows in length and having a more extensive surge component as evidenced by the greater zone of singed vegetation around the main flow body. This flow was triggered by preceding major collapse events which removed significant amounts of the "cap" over the pressurized conduit, allowing a sudden release of pressure (Yamamoto et al. 1993). Several other vulcanian explosions occurred in the summer of 1991, but none were accompanied by pyroclastic flows. The explosion on 11 June was the most powerful of these and resulted in pumice up to nearly half a meter in diameter falling as far as 5 km to the NE of the crater in the populated Senbongi district of Shimabara, resulting in the breakage of numerous car windscreens. About 20,000 tonnes of volcanic bombs are thought to have been ejected by this explosion alone.

Pyroclastic flows initially proceeded to the east along the Mizunashi River valley (Yamamoto et al. 1993). The distances they traveled gradually increased, reaching 3 km on May 26, 1991, 3.2 km on 3 June (accompanied by a fatal 4-km-long surge—details below), and 5.5 km on 8 June (the longest runout of the whole eruption). Dome growth temporarily shifted toward the NE flank in the following months. Consequently, PFs started to flow down toward the Taruki

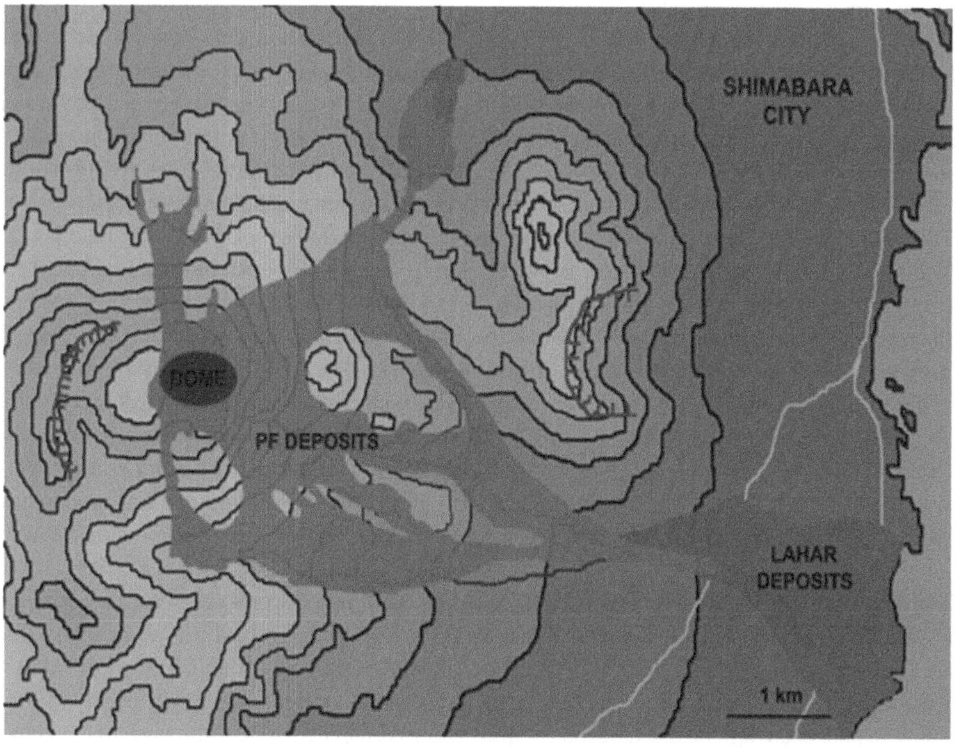

Fig. 12.2 Cumulative Pyroclastic flows and lahar deposits, 1990–1995 Eruption. Source: http://www.photovolcanica.com/VolcanoInfo/Unzen/UnzenTotal.jpg

Height before turning right into the narrow Oshigadani River gorge which feeds into the Mizunashi River about 3km from the dome. On September 15, 1991, a large collapse event led to an extensive flow along this drainage, the surge cloud which crossed the path of Mizunashi River and impacted infrastructure, including the Onokoba Elementary School. This was the most voluminous flow of the whole eruption. A further notable flow occurred on June 23, 1993. This was one of several flows that entered the Senbongi residential district after crossing the Taruki Height NE of the volcano and was responsible for one fatality (a local viewing his house that had been burnt during a previous flow). Senbongi had become exposed to PFs since previous activity had flattened the topography by filling the Oshigadani valley with deposits.

12.3.3 The Impact of the Eruption

The impact on settlements and economic activity from the eruption included the burning of 800 buildings by PFs and the destruction of 1700 houses by debris flows/lahars. A total of 44 fatalities were recorded, and over 11,000 people were displaced at the height of the eruption when the restricted areas were widened as an immediate response to the fatal 3rd June 1991 event. This event gained fame in the geological community due to the deaths of two of the world's most noted geo-event filmmakers, the French couple Katia and Maurice Krafft, a US Geologist, and 40 Japanese casualties during its progress. On June 3rd, 1991, the Krafft's entered the restricted zone around the volcano to film the activity. The position they occupied was in the direct line of pyroclastic flows proceeding down the Mizunashi River, which however took a slight turn to the right before their position. Previous flows had reached a maximum distance of about 3 km a week earlier, but had yet to reach as far as the position they chose.

In the afternoon (at 16:08 local time), the eastern half of the lava dome collapsed, together with some older underlying material. About 0.5 million cubic meters of material was mobilized and rushed down the Mizunashi drainage in the form of a massive Pyroclastic Flow, far surpassing all prior events during the eruption (Fig. 12.3). The PF body reached as far as 3.2 km, but the highly energetic ash surges accompanying it swept on, reaching a maximum distance of 4 km by about 16:10. The flows were so powerful that cars of some of the journalists were swept away, one being moved as far as 80 m. Trees were flattened in the area as were most of the houses. All people in the area were killed by the impact of the surge and/or by the intense heat. The ash surges may have been particularly intense as the result of the flow cascading over several waterfalls as it went, thus increasing fragmentation and production of the fluidization zone overlying the main flow body. A fatal event at Mt. Merapi (Indonesia) in 1994 has similarly been attributed to such a phenomenon (Erfurt-Cooper 2014).

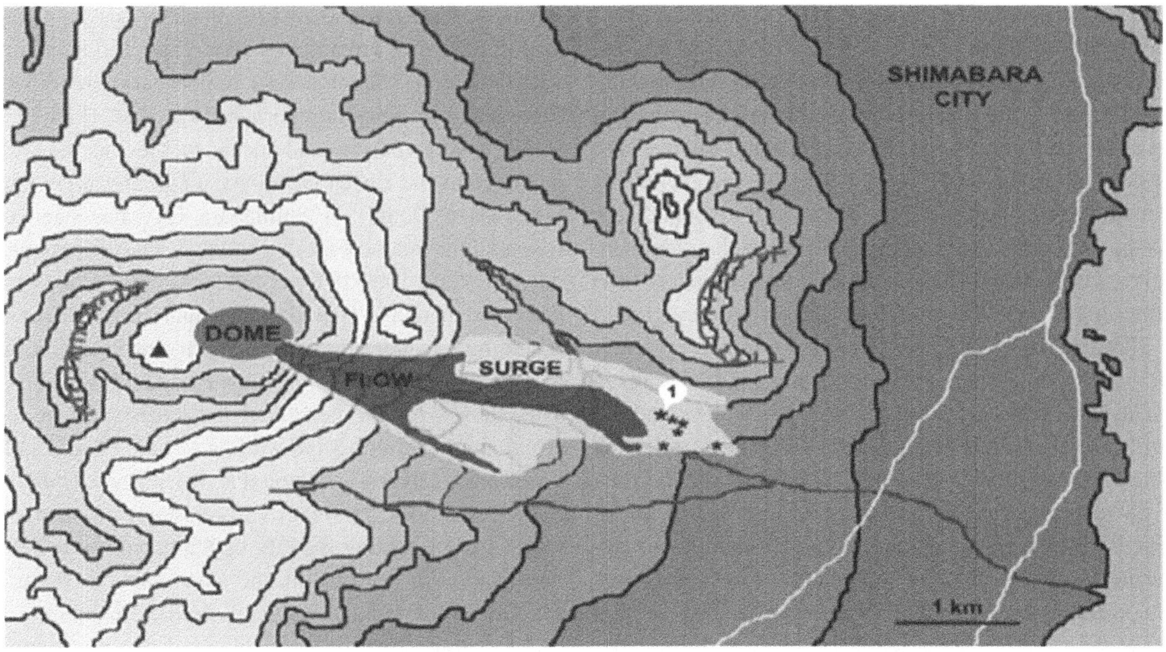

Fig. 12.3 The Fatal Pyroclastic Flow of June 3, 1991. Source: http://www.photovolcanica.com/VolcanoInfo/Unzen/UnzenDeadly.jpg. Note: Sketch map showing approximate extent of fatal pyroclastic flow and surge on June 3, 1991. Stars indicate victim locations. Star labeled (1) is the site where the bodies of Maurice and Katia Krafft were found

12.3.4 Later Flows

The pyroclastic flows of September 15, 1991, included the longest flow passing through the Oshigadani drainage channel during the entire eruption and involved the mobilization of about 2.4 million cubic meters of dome material. The events on this day have been analyzed in detail by Fujii and Nakada (1999). The main flow initially descended the NE flank and was then deflected by the Taruki height, directing it in a SE orientation through a narrow gorge, after which it flowed into the lower reaches of Mizunashi River, which channeled the flow eastwards toward the coast. As the flow turned at the Mizunashi River, a large surge detached and continued in a SE direction for another 600 m. This surge set light to about 150 houses in the Ohnokoba district as well as the local elementary school, which remains as an outdoor natural disaster exhibit today. The destructive nature of the flow and the large volume of the associated surge, especially as it exited the Oshigadani gorge, can possibly be attributed to increased collision of materials in the flow body as the flow passed through the gorge, resulting in generation of more fine materials.

Progressive failure of the NE flank of the lava dome started at 16:44 local, with further major events following at 17:59 and 18:45, before the climactic flow began at 18:54 (Fujii and Nakada 1999). The first flow stopped just north of Kita-Kamikoba, while the next two preceding the third or main flow were slightly shorter. This event was far more energetic and surge clouds flanking the main flow body flattened vegetation on either sides of the flow and even remobilized trees already toppled by the fatal June 3 event in the Kita-Kamikoba area, while also transporting the remains of a car about 120m. As the surge subsequently detached from the flow, it rapidly lost energy but remained dense and extremely hot. At the school, vinyl pipes of the north side (facing the surge) were melted, but most windows withstood the de-energized surge. Interestingly, about 20 cm of ash was deposited in the playground and on this side more windows were broken. Hot ash appears to have entered the building and set the classrooms alight on the playground side. The deposits of the flow have been studied in detail, and it was reported that they generally contain three units: a 20 cm layer of well-sorted ash, up to 2 m thick block and ash layer rich in fresh, slightly vesiculated dacite lava, and a few cm thick layer of ash which would have settled on the flow after its passage. The bottom layer of well-sorted ash is attributed to the surge running ahead of the main flow body and being overrun by the flow immediately after deposition. This correlates to video footage showing highly mobile lobes (of the surge) "jetting ahead" at the base of the main flow body.

12.3.5 Mt. Unzen in the Wider Context

In a reflection of how the geological heritage of a site like Mt. Unzen is important to local development and our understanding of natural capital, some of these disaster zones described earlier were later developed into major tourist attractions. The "Buried Village," for example, is part of a suburb of Shimabara City and is a constant reminder of the dangers of living close to active volcanoes. This area was in the flow path of the lahars which followed the pyroclastic flows, with coincidental monsoonal rains causing large mudflows that destroyed over 2500 houses. Every day busloads of tourists bring domestic and international visitors to view the destruction caused (Figs. 12.4, 12.5, and 12.7), and these tourists often take the opportunity to talk to residents about their ordeals during that time. Tourists also go to other disaster sites like the primary school that was severely damaged by a pyroclastic flow.

The volcano itself and its surrounds are covered in a set of "GEO SARAKU" model course recommendations. Information on these car-based tourist routes can be accessed through the Unzen Volcanic Area Global Geopark Office in the Disaster Memorial Hall. The first is an approximately 4-h experience of the course of evolution of the volcanic area from 4.3 Ma; the second retraces the "Shimabara Catastrophe" (the worst volcanic disaster in Japan's recorded history and takes 3 h. Other courses cover the hot springs and Unzen Jigoku ("hells," or eruption points) in the area of the volcano and the battlefields/middle ages iron-making/castle ruins in the area that introduce a historical social ambience to the Geopark. Access to the area is gained by bus from Nagasaki airport (90 min), Fukuoka by train (3 h), and Kumamoto by ferry (30 min). Other access options are car (3.5 h from Fukuoka) and subsidiary ports/railway lines (Fig. 12.6).

To enable tourists and locals to see the most chilling evidence of the Unzen eruptions, a *Disaster Memorial Hall* (Gamadasu Dome) in Shimabara City has been created (open all the year except during the 7–11 April maintenance days; fees are 1000 JPY per adult, with reductions for children and seniors; free parking) and has many exhibits, including a camera found in the pyroclastic debris that still had a few seconds of footage that could be saved (Fig. 12.7). This shows, for the benefit of the tourists, the pyroclastic cloud approaching people waiting to take photos and make video documentaries. As noted above, 43 people died in this pyroclastic flow. However, this has not stopped volcano tourists from visiting the city of Shimabara to benefit from the many hot springs in the area, and to drive up to Mt. Unzen for walks and picnics on the mountain.

Next to the destroyed school, a volcano observatory was built which also includes a commemorative information

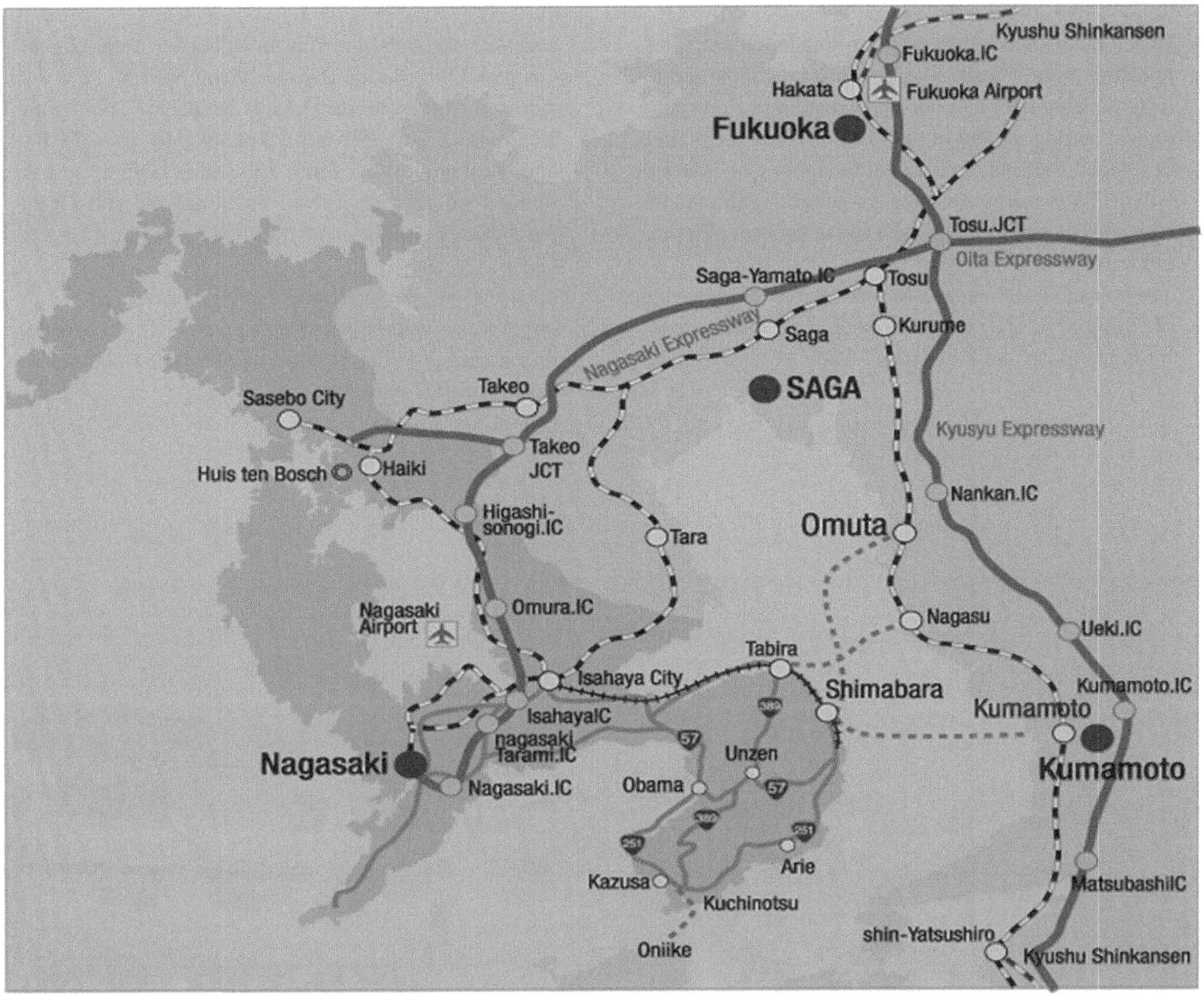

Fig. 12.4 Access to the Geopark. Source: Courtesy of the Unzen Volcanic Area Global Geopark Japan

center, a museum, and a volcano viewing platform for visitors. Another sightseeing location popular for viewing Mt. Unzen is from a lookout at the Nita Pass on the flanks of the volcano, again frequented by busloads of visitors who cannot get *close enough* to an active volcano, especially one with such a reputation.

Heritage in a Nutshell: Unzen Geopark (The city of Shimabara and its Surrounding Landscapes)
Type of Heritage: Composite: Geological-Geomorphological-Community Heritage Complex involving active volcanoes (classified as a Decade Volcano); Global Geopark designation in 2009; as well as a Dark Tourism connection through the devastation wrought by the 1991–1995 eruption in and around the city of Shimabara. Decade volcanoes are a set of 16 identified by the IAVCEI as requiring constant attention in the "light of their history of large, destructive eruptions, and their proximity to populated areas."
Mount Unzen is an active composite stratovolcano in the Southwest Japan Arc. It is in fact a cluster of overlapping volcanoes that have formed at various times over the past 6 million years in the Shimabara Peninsula. There have been many disastrous explosive eruptions from this volcano, and it is currently active. Its most recent eruption began in 1991 and lasted until 1995. Unzen-dake got its name from all the hot springs found in its vicinity. Originally, the mountain is said to have been called "Onsen-dake," onsen means hot springs, and "dake" means peak, so the translation of

(continued)

Unzen-dake into English is "hot spring mountain." In a reflection of how a site like Mt. Unzen is important to local development and our understanding of our geological heritage, some of the disaster zones were later developed into major tourist attractions. The "Buried Village," for example, is part of a suburb of Shimabara City and is a constant reminder of the dangers of living close to active volcanoes.

The volcano itself and its surrounds are covered in a set of "GEO SARAKU" model course recommendations. Information on these car-based tourist routes can be accessed through the Unzen Volcanic Area Global Geopark Office in the Disaster Memorial Hall.

Access: Access to the area is gained by bus from Nagasaki airport (90 min), Fukuoka by train (3 h), and Kumamoto by ferry (30 min). Other access options are car (3.5 h from Fukuoka) and subsidiary ports/railway lines.

Useful websites:
https://www.volcanodiscovery.com/unzen.html; http://www.unzen-geopark.jp/en-first; http://www.unzen.org/e_ver/faq/; http://vivaweb2.bosai.go.jp/v-hazard/pdf/12E.pdf

Fig. 12.5 Tourism at one of the disaster zones in Shimabara. Source: Photo courtesy of the Author. Note: The roof of one of the many houses buried by lahar flows is visible in the center of the picture. This is a tourist attraction and one of the must-see stops when busloads of visitors come to Shimabara

Fig. 12.6 Local communities are intensely involved in disaster prevention and share their experiences with tourists. Source: Photo courtesy of the Author

Fig. 12.7 One of the cameras found after the eruption of Mt. Unzen Source: Photo courtesy of the Author

12.4 Conclusions

Despite the continuous activity found at many of the volcanoes of Japan, all of them including Mt. Unzen are also significant attractions for domestic tourists and are increasingly important for international visitors. While the attractions and activities in Japan are of course the same as those in similar areas in other countries, including both geosites and adventure tourism as they do, they also have one other major attraction as items of natural heritage: millions of Japanese people see them as the providers of the natural hot mineral water that plays a big role in the social life of this country. Thus, the tourism infrastructure in active geosites generally includes the Onsen (hot springs), related accommodation, and historical sites and restaurants, while the natural heritage itself is very much valued as an *everyday* commodity.

While the Mt. Unzen area as a decade volcano is a prime example of how extreme such geosites can be, it is also a very important natural heritage item that can aid in the understanding of volcanic geological phenomena from both the geoscience and community points of view. In addition, Japan is very well situated when it comes to dealing with safety issues in public places (Cooper and Erfurt 2007), so the natural heritage of Japan is also a very important educative tool and site for the understanding of geological risk management. Through the events of 1990–1995 at Mt. Unzen, there has been a growth in understanding of the geological environment's impact on human settlements and the related costs of these impacts. We all need to understand the importance of our geological heritage, and the increasing importance of conserving it in all countries, if we are to use this heritage for our own and future generations' natural capital.

References

Aramaki S, Ui Y (1982) Japan. In: Thorpe RS (ed) Andesites. Wiley, New York, pp 259–292

Cooper MJ (2014) Volcanic National Parks in Japan. In: Erfurt-Cooper PJ (ed) Volcanic tourist destinations. Springer, Heidelberg, pp 231–246

Cooper MJ, Erfurt PJ (2007) Tsunamis, Earthquakes, volcanism and other problems: disasters, responses and Japanese tourism. In: Laws E, Prideaux B, Chon K (eds) Crisis management and tourism. CABI, Wallingford, pp 234–251

Dowling RK, Newsome D (2010) Global geotourism perspectives. Goodfellow Publishers, Oxford

Erfurt-Cooper PJ (ed) (2014) Volcanic tourist destinations. Springer, Berlin

Erfurt-Cooper PJ, Cooper M (2010) Volcano and geothermal tourism: sustainable geo-resources for leisure and recreation. Earthscan, London

Fujii T, Nakada S (1999) The 15-September 1991 pyroclastic flows at Unzen Volcano (Japan) a flowing model. J Volcanol Geotherm Res 89:159–172

Hoshizumi H, Nakano S (2009) Active volcanoes and volcanic disasters. National Institute of Advanced Industrial Science and Technology, Tokyo

Hoshizumi H, Uto K, Watanabe K (1999) Geology and eruptive history of Unzen Volcano, Shimabara Peninsula, Kyushu, SW Japan. J Volcanol Geotherm Res 89(1–4):81–94

Hoshizumi H, Uto K, Matsumoto A, Kurihara A (2004) Growth history of Unzen Volcano, Kyūshū, Japan, Fall Meeting. American Geophysical Union, San Francisco

Japan Meteorological Society (JMA) (2007) Volcanic activity level. JMA, Tokyo

Johnston E (2011) Latest Volcano Show: Shinmoe. Japan Times, 1 March, p 3

King L, Brattstrom T (2014) Halema'uma'u's Lava Lake Brightens after Dark Geotourism in Hawai'i Volcanoes National Park. In: Erfurt-Cooper PJ (ed) Volcanic tourist destinations. Springer, Berlin, pp 283–294

Kuno H (1962) Part XI, Japan, Taiwan and Marianas: catalogue of the active volcanoes of the world including Solfatara fields. International Association of Volcanology, Rome

Machida H (2015) Outline of tectonic setting and explosive volcanism of southern Kyushu. In: Moriwaki H, Nagasako T, and Okuno M (eds) Excursion guides: Post-congress excursion PO-13, Quaternary volcanism, tephras and associated landscapes in Kyushu Island. XIX INQUA Congress, pp 32–56

McKeever PJ (2009) UNESCO Global Network of National Geoparks: geological heritage and sustainability. Department of Enterprise, Trade and Investment, Geological Survey of Northern Island, Belfast

Miyabuchi Y (1999) Deposits Associated with the 1990–1995 eruption of Unzen Volcano, Japan. J Volcanol Geotherm Res 89 (1–4):139–158

Nakada S, Shimizu H, Ohta K (1999) Overview of the 1990–1995 eruption at Unzen Volcano. J Volcanol Geotherm Res 89(1–4):1–22

Siebert L, Simkin T. (2002) Volcanoes of the World: an illustrated catalogue of holocene volcanoes and their eruptions. Smithsonian, Global Volcanism Program Digital Information Series, GVP-3, Washington, DC

Smithsonian Institute (2015) Global Volcanism Program. Natural Museum of Natural History, Smithsonian Institute, Washington, DC

Tilling RI (1989) Volcanic hazards and their mitigation: progress and problems. Rev Geophys 27:237–269

Uto K, Hoshizumi H, Matsumoto A, Oguri K, Nguyen H (2001) Volcanotectonic history of Shimabara Peninsula and the evolution of Unzen volcano in Southwest Japan, Fall Meeting. American Geophysical Union

Yamamoto T, Takarada S, Suto S (1993) Pyroclastic flows from the 1991 eruption of Unzen Volcano, Japan. Bull Volcanol 55(3):166

Challenges for Geoconservation in Contemporary Japan 13

Abhik Chakraborty and Kuniyasu Mokudai

Abstract

This chapter analyzes the challenges for geoconservation in Japan, based on the findings of a questionnaire survey. Geoconservation is defined as the action taken with the intent of conserving earth heritage, but it has proven difficult to implement this concept at the ground level due to perceived low priority of conservation and lack of funding. In the case of Japan, the predominance of the natural hazards discourse has resulted in extensive modification of watersheds and coastlines by engineering, affecting the surface level features of geodiversity. Although there are a large number of national geoparks and eight UNESCO Global geoparks in Japan, the study found that those geoparks currently lack substantive information on the anthropogenic threats on geodiversity, a robust monitoring scheme to track down change, and adequate expertise to mitigate fragmentation and possible loss of geodiversity. The study also concludes that while international recognition such as the UNESCO Global Geopark brand does not bring additional protection, such recognitions do seem to elevate stakeholder awareness for geoconservation.

Keywords

Geoconservation • Geodiversity • Fragmentation • Geoconservation Audit • Monitoring

13.1 Geoconservation: A Dynamic Concept, but a Difficult One to Ground

Geoconservation is defined as "action taken with the intent of conserving and enhancing geological, geomorphological and soil features, processes, sites and specimens, including associated promotional and awareness raising activities and the recording and rescue of data or specimens from features and sites threatened with loss or damage" (Prosser 2013). However, as Prosser (2013) notes, practices of geoconservation can differ according to countries and regions and according to the maturity and perceived importance of the concept by relevant actors. In this regard, Japan is still at an early stage of conserving its geological and geomorphological heritage, although specific "landmarks" (landforms) are perceived as important assets for tourism and for earth science education. The situation is in contrast to European countries, notably places such as the UK, which have built a track record of geoconservation and geodiversity evaluation over the years. It is noted by Prosser (2013) that in the UK, national geoconservation strategies date back to the year 1949 with the introduction of the national statutory system of nature conservation (eventually leading to designation of thousands of Sites of Special Scientific Interest) (Whiteley and Browne 2013), and Burek and Prosser (2008) provide a detailed description of geoconservation initiatives of the twentieth

Electronic supplementary material: The online version of this chapter (doi:10.1007/978-3-319-61896-8_13) contains supplementary material, which is available to authorized users.

A. Chakraborty (✉)
Center for Tourism Research (CTR), Wakayama University, Wakayama, Japan
e-mail: abhik78@hotmail.co.jp

K. Mokudai
Japan Geo Service Inc., Tokyo, Japan
e-mail: mokudai@japangeoservice.com

A. Chakraborty et al. (eds.), *Natural Heritage of Japan*, Geoheritage, Geoparks and Geotourism,
DOI 10.1007/978-3-319-61896-8_13

century. However, it should also be noted that the concepts of geodiversity and geoconservation were formally developed much later. The seminal works of Christopherson (1992) and Gray (2004, 2013) mark important milestones of drawing scientific and management attention to the concepts of geosystems and geodiversity, respectively. In this regard, it is apt to consider the period after the 1990s as the real phase of geoconservation (or geodiversity-conscious conservation). Two major developments that took place in the UK at this time were the establishment of voluntary geoconservation groups (with autonomous funding in many cases) (Prosser 2013) and the Regionally Important Geological or Geomorphological Sites (RIGS) scheme (Whiteley and Browne 2013), both of which can be considered examples of autonomous, small-scale, and bottom-up approaches toward evaluation and conservation of geodiversity. Such local level efforts are complemented by schemes such as the Geological Conservation Review (Ellis 2011) and geodiversity forums such as the Scottish Geodiversity Forum (Scottish Geodiversity Forum nd) that have made national and subnational level audits of geodiversity and conservation possible. Similar examples can be provided from Australia [The Australian Natural Heritage Charter, see Gray (2013)], Spain (Garcia-Cortes et al. 2012), and New Zealand (Kenny and Hayward 1993). Toward the end of this chapter, we would revisit this important point to propose some possible pathways for geoconservation in Japan.

Another important development for geconservation is observed to be the birth of the Global Geoparks concept (initially promoted by the Global Geoparks Network but currently run as an official UNESCO program) (Larwood et al. 2013). Although the Geoparks and the UNESCO World Heritage programs have recognized important geodiversity-related features, as Erikstad (2013) points out, these international brands do not bring any additional legal protection for sites in threat. Matthews (2014) pointed out that in the European context, although there has been a rise in the geopark numbers, those properties remain isolated and large sections of the European geodiversity remain fragmented. Worton and Gillard (2013) observe that those involved in geoconservation in the UK often feel "undervalued" compared to professionals engaged in other disciplines. Therefore, although international programs like the Global Geoparks possess excellent potential for promoting geoconservation, a mere growth of their numbers may not provide an accurate estimate of geoconservation in the concerned area. Erikstad (2013) also points out that as geoconservation is based on the concept of geodiversity (i.e., the "diversity" needs to be conserved, not specific landmarks or artifacts that have economic potential), its challenges are not amenable to simplistic solutions. In relation to these observations, it can be noted that geoconservation has remained a difficult concept to be grounded.

How the geopark designation movement spread from Europe to Japan, eventually leading to a large number of national geoparks (and eight UNESCO Global Geoparks) being established within a relatively short phase, is described by Watanabe (2017). We have been involved with analysis of geoconservation challenges and geopark evaluation of Japan for some time, and in 2015, we organized an international session in the annual Japan Geoscience Union (JpGU) convention, which was possibly the first session of its kind in a major academic conference in Japan. Later in 2015, we carried out a questionnaire survey with the help of the Japanese Geoparks Network to gauge the level of understanding and current problems of geoconservation. Below, we report on the findings and discuss potential improvements.

13.2 A Note on Japan's Geodiversity Background and Main Challenges

Japan's geodiversity environment is shaped by the location of these islands in an active continental margin. Wakita (2017) describes the long geological history of the formation and evolution of these islands and corresponding observable features. Neogene and Quaternary volcanism, subduction, and island arc formation processes—i.e., the most recent phase of the geological history of the archipelago—attract most attention. Volcanism as a process that results in significant ongoing formation of geodiversity is discussed by Nakada (2017). Due to its location in the Asian Monsoon Belt, particularly vigorous denudation is observed in the Japanese Islands along with rapid uplift and crustal deformation, a situation that led to the term "Tectonically Active and Intensely Denuded Region" being applied to this region (Yoshikawa 1984). One of us (Mokudai 2014) has discussed how past "natural hazards" have shaped the "natural bounties" such as relatively flat lands and alluvial plains of today, and how the society coevolved with these geological forces.

While the Japanese Islands have a rich and diverse geological and geomorphological history, a look at the themes of the national geoparks of Japan (for details, see http://geopark.jp, http://geopark.jp/en/geopark/) reveals that most geopark activities are concentrated on interpreting volcanic eruption events, natural hazards, and hazard prevention schemes. In addition, most featured contents of geopark activities are predominated by "showcasing" geopark attractions (landforms, landmarks, cultural interpretations) and tourism promotion or basic earth science education-related information. There is little substantial information on threats to geosites or geodiversity, possible mitigation schemes, and utlilization of geoparks as tools to raise awareness on local or global environmental problems. From the contents currently featured on the individual geopark homepages, a systematic and functional evaluation of geodiversity and conservation of that diversity (or even conservation details for most important geological landforms) is not available either. Although many landmarks fall under the existing National Park areas (subject to relevant laws), or are covered by the Cultural Properties Conservation Law, neither law has specific provisions for geoconservation. In fact, we noted serious threats of fragmentation and possible ongoing loss of geodiversity on the ground in some geoparks (Chakraborty et al. 2015; Chakraborty and Mokudai 2016).

The situation is compounded by the extensive development that has obscured surface level geological and geomorphological features (diversity) and led to an increase of homogenous, mostly concretized, land cover. It is estimated that only 45% of the total length of coasts in the four main islands of Hokkaido, Honshu, Shikoku, and Kyushu are in a relatively unmodified state (Knight 2010). Rivers, a most important geomorphological agent, are compromised in their functions due to extensive damming, embankment fortifications, artificial channelization, and flow regulation (Uzawa and Okuma 2010; Graaf and Hooimeijer 2008). While in many cases some trade-offs are inevitable and even necessary, the overall condition poses significant challenges for appreciation of geodiversity and for the integrity of the processes that lead to the formation and evolution of geodiversity—leading to obvious challenges for geoconservation.

13.3 Data Collection and Analysis Method

A questionnaire was distributed to all registered Japanese Geopark Network members (current UNESCO Global, national, and aspiring national geoparks in Japan), and responses were collected through emails. Valid responses were collated and analyzed through simple descriptive statistics. As the valid responses are 43 in total, they correspond to 78% of the then registered members (55) and 75% of members (57) when this result is being published. Therefore, the results remain valid and are highly representative of the situation in Japan. The data was collated twice, once for the whole group of 43 geoparks and aspiring regions (these include UNESCO Global Geoparks, Japanese national geoparks, and aspiring Japanese geoparks) and separately for the eight UNESCO Global Geoparks (all of which responded). The survey was conducted in Japanese; results have been translated into English.

13.4 Results of the Geoconservation Survey

The results of the Geoconservation Survey are elaborated below. For more details, refer to the "Extra Supplementary Material (ESM)" to this chapter.

First, geoparks were asked how they interpret the word "nature" (What do you feature as "nature" in your geopark?). Multiple answers were possible for this question. Out of the 43 respondents, 34 (79.1%) selected "geology," 22 (51.2%) selected "fossils," 40 (93%) selected "landforms and landscapes," 40 (93%) respondents selected "forests, grasslands, satoyama production landscapes," 37 (86%) selected "insects, birds, and mammals" and 42 (97.7%) selected"rivers, lakes, and sea (including coastal environment)." The responses reveal that the majority of the geoparks possess a correct and varied understanding of "nature." However, 8 (18.6%) replied that"archaeological remains," 2 (4.7%) replied that "town/city scape," and 7 (16.3%) replied that "mines" are interpreted as "nature." These three options were decoys. Though statistically small, these numbers indicate that in some geoparks cultural artifacts, modern housings, and extractive activities are erroneously seen as parts of "nature." When results were collated for the 8 UNESCO Global Geoparks, the erroneous responses fell somewhat, nobody selected "town/cityscapes," but 2 (25%) still selected "archaeological remains," and 1 (12.5%) selected the "mines" decoy. This shows that the concept of "nature" and natural diversity is still not fully grounded in even UNESCO Global Geoparks.

Next, respondents were asked about the "condition" of the main geological features such as strata, outcrops, rocks, and minerals. While 21 (48.8%) replied that those features are currently protected by "law," 9 (20.9%) replied that those are currently affected by collection and extraction, and 6 (14%) replied that those are affected by construction. For UNESCO Global Geoparks, the protected condition extends to 75%, and only 12.5% of signature geological assets are currently affected by collection or extraction. This shows a markedly different situation as far as protection levels between UNESCO geoparks and national geoparks are concerned. If the "affected" categories are amalgamated for the total respondents group, we can see that nearly 35% of these features are currently facing threats. This allows us to draw two tentative conclusions. One is that a significant part of such geodiversity is currently under threat, and the other is that although UNESCO geoparks supposedly do not bring additional legal stipulations for protection, UNESCO geopark status possibly creates enough "peer pressure" or "momentum" for additional preservation measures.

When asked specifically about the fossils, out of a total of 20 responses where it was applicable 6 (14%) replied that the fossils were either wholly or nearly fully conserved in situ. Out of 4 relevant UNESCO Global geopark responses, only 1 (12.5%) replied of in situ conservation of fossils. This outcome is somewhat surprising as it runs counter to the finding from the previous question. When asked about the condition of forests, grasslands, or managed forests, 60.5% (26 respondents) replied that natural forests occupied a large fraction of total forested land and that those areas are mostly protected. The relevant response from UNESCO Global Geoparks shows a slight fall to 50% (4 properties). In contrast, when asked about rivers, wetlands, and coastal environment, 26 (60.5%) and 14 (32.6%) replied that rivers and coastal areas were "substantially changed" due to engineering, respectively.

For UNESCO Global geoparks, 25% and 37.5% responses indicated similar change to the relevant categories. The response outcome was expected in the light of the overall situation of intensive engineering on watercourses in Japan. In what stands out as a positive note, 51.2% of all respondents and 62.5% of UNESCO geopark respondents identified "natural sections" in their coastal environments.

Next, in order to gauge the priorities of geopark activities, respondents were given choices to determine "why" they chose geopark promotion for their area. Whereas 62.8% (27 respondents) chose "Geoparks are places for both abiotic and biotic nature and there are ongoing species habitat conservation schemes in our area," 23.3% (10 respondents) chose the option "The geopark concept is anthropocentric and is mainly based on utilizing geological features and our area is promoting geoparks mainly for economic reason." Even 1 UNESCO global geopark chose the same response. This situation shows a clear contrast, and as we would like to point out in the discussion section, the relevant national committees should carefully select geoparks and examine the priorities for the concerned areas. When asked about the frequency of lectures, workshops, or events related to nature conservation, only 4 (9.3%) of the total respondents could reply that they did so for multiple times a year, and 2 of those were current UNESCO global geoparks. 30.2% of respondents did not call on a nature conservation expert for their educational events, and 16.3% had no experience of running such events. From these responses, it becomes clear that involvement of nature conservation experts remains low and there is much room for improvement.

The situation is even more serious as far as monitoring of the geodiversity and the natural environment is concerned. Only 3 respondent areas out of 43 (7%) feature active monitoring programs run solely by the relevant geopark promotion and management bodies; in case of global geoparks, only 1 geopark could choose this answer. Although responses reveal that some form of monitoring exists, either by the Ministry of the Environment or "voluntary" monitoring by geoguides (32.6% and 30.2%, respectively), 41.9% (18 areas) also chose "we do not have ongoing monitoring," which is a significant ratio. If UNESCO geoparks are chosen, 75% show some monitoring in liaison with a relevant ministry and 62.5% show existence of voluntary monitoring. This result again confirms that the UNESCO accreditation creates positive pressure for nature conservation. When asked whether the geoparks have any qualified conservation experts (degree PhD or above in a relevant discipline), only 2 geoparks (4.7%) could reply in the affirmative, and surprisingly, no UNESCO global geopark had that answer, and 25 respondents (58.1%) chose an overall negative answer (no specialist within geopark, no liaison, no outsourcing, no advice from outside). This clearly shows that geoparks in Japan suffer from a serious lack of nature conservation (and therefore geoconservation) experts working for them; such a situation poses a serious challenge for the conservation of geodiversity. As for any formal tie-up with a nature conservation organization (including nonprofit groups), only 4 (9.3%) territories could reply in the affirmative. However, it also became clear that UNESCO geoparks have a higher percentage of informal interaction with nature conservation groups (62.5) compared to the percentage of all geoparks combined (34.9).

Next, respondents were asked about the level of information they possessed about threats on their natural environment and whether geoparks seek to restrict activities that are possibly damaging for geodiversity conservation. Twenty-seven territories (62.8%) had no Red List (IUCN endangered species information), and no single respondent indicated any experience of doing surveys with the local population on the threats posed to their natural environment (geodiversity) (42 respondents replied in the negative, while 1 did not answer). This finding clearly shows that surveys on the threats on geodiversity are practically nonexistent, even in UNESCO global geopark territories. One important factor behind this situation is the availability of funding. Thirty-three territories (76.7%) replied that they have no funding for such surveys. Even a majority of the current UNESCO global geoparks (5 territories) are afflicted by the same problem. Similarly, 79.1% (34 territories) had no mandatory EIA for development projects and only 4 territories (a mere 9.3%) have a conservation plan (3 out of those are UNESCO Global Geoparks). In what should be a matter of concern, 19 territories (44.2%) replied that they neither have a conservation plan nor do they plan for making one for their territory. This is consistent with 46.5% (20 territories) of all respondents and 62.5% (5 territories) of global geoparks replying that even local residents do not have "adequate understanding" of the need of conservation. Asked what they would like to do in order to improve the current situation and contribute to nature conservation, 23.3% chose formulation of a conservation plan as a priority, and 39.5% indicated that they would seek further understanding from the local community and 39.5% replied that they would like to organize a related event in their geopark (multiple answers were possible). If current global geoparks are singled out, 37.5% prioritized a conservation plan and 50% prioritized increased interaction with the local community.

13.5 Discussion

From the survey, three main problems could be identified as far as challenges for geoconservation are concerned:

1. Lack of any systematic evaluation of threats posed to geodiversity and geoconservation audits

2. Lack of experts and planning mechanisms for geoconservation and lack of funding for achieving the same
3. Perceived "low importance" for geoconservation, not only in the management level but also possibly in the local society.

Some positive signals could be identified: the high ratio of "natural" and protected forest cover, existing legal protection of important geological features, and the willingness to consider nature conservation (geoconservation) as part of geoparks are noteworthy.

From this survey, it became clear that no geopark undertook a participatory survey on geodiversity conservation with the local residents and that currently the involvement of nature conservation professionals (as either geopark workers or advisers) remains at a low level. In addition, very few geoparks have formal tie-ups with organizations or groups devoted to nature conservation. The result is the corresponding lack of awareness in both geopark management and (possibly) local society levels about the threats posed to geodiversity of the concerned areas. From the questionnaire, it also became clear that nearly 35% of the important geological landmarks and landforms are currently under some sort of "threat." This is a significant ratio of potentially damaged geodiversity. If we consider geomorphological processes at the landscape level such as watersheds and coastal belts, the threat is probably even more serious, as substantial recent alterations of these environments were reported. As for fossil collections, although these resources are conserved, in situ conservation ratio is low (14%). again possibly implying ongoing disturbances to the areas of fossil or mineral deposits. As very few geoparks have made EIA mandatory for development projects, such disturbances at the landscape level are likely to continue, unless the relevant management bodies seriously consider about assessing damage to sites and make EIA mandatory in geopark areas. A related problem is that most Japanese geoparks are composed of multiple administrative units including urban and industrial areas, and implementation of rigorous EIA might not find consensus among the stakeholders. Lack of funding, as observed above, remains a serious issue. Although funding-related challenges are posed to geoconservation groups worldwide, it was noted, with reference to Prosser (2013) and Whiteley and Browne (2013), how local level groups have managed to obtain funding for geodiversity assessment and geoconservation schemes in Europe, leading to successful geoconservation efforts in places such as the UK.

Two related challenges are: the current emphasis toward "showcasing" geological or geomorphological features rather than utilizing the geopark as a forum for understanding and protecting natural diversity and the close similarity of themes in most Japanese geoparks. As far as the first challenge is concerned, the situation is compounded by the lack of "monitoring" the natural environment. Geoparks currently do not possess enough information on the changes and possible ongoing deterioration in the natural environment (diversity) (a fact underlined by their lack of information on endangered species), nor do they have the necessary expertise (lack of professionals involved) to address this. As Japanese geoparks are overseen by the Japan Geopark Committee (JGC), an independent body that has scientists as members, it could, in future, contribute to geoconservation by holding geoparks accountable for ongoing fragmentation or damage to geodiversity and by offering skilled advice. As far as the second challenge is concerned, the number of total geoparks is already quite high (43 national geoparks including the UNESCO geoparks, and 14 aspiring geoparks as of March 2017), and in a region shaped by a dominant geological history (formation and evolution of island arcs), theme overlaps are bound to occur. Currently, volcanism–subduction–crustal deformation and natural hazards related themes predominate, and it is possible that due to the strong correlation between active earth processes and life and property damage, the geoconservation discourse has not gained maturity so far. The JGC should carefully select regions that can offer truly outstanding features of the long geohistory of the Japanese Islands backed up by active and responsible conservation of this heritage.

A third important lesson was gained from the survey: international branding such as global geoparks seems to result in greater responsibility toward geoconservation, and at least, such branding leads to a heightened perception about the importance of geodiversity. This is reflected in the higher percentage (62.5) of UNESCO geoparks having informal working relationships with nature conservation groups than the combined group of 43 geoparks (34.9). Therefore, we conclude that the UNESCO accreditation does bring some geconservation benefits, albeit in an indirect manner. As there are 8 UNESCO geoparks in Japan, it could be foreseen that appreciation of geodiversity and the sense for protecting it will be felt in a stronger manner in these areas.

13.6 Some Proposals for Improvement

While the current vista of geoconservation is full of challenges, there are some positive signs and potential for successful grounding of geoconservation in the future. And while some of the challenges such as lack of funding or lack of environmental impact assessment might take a long time to be addressed, there are some problems that require urgent attention in order to protect and promote the geoheritage of the Japanese Islands. Geoparks—especially UNESCO Global Geoparks—have clear responsibilities toward this, as outlined in the statutes of the International Geoscience and Geoparks Programme (UNESCO nd). Below we mention a few ideas for improvement of the current situation.

Firstly, geoparks are affected by lack of qualified nature conservation professionals working for them. Either geoparks should consider employing nature conservation professionals with proven track records, or they should consider outsourcing such tasks to qualified professionals and strengthen tie-ups with relevant national organs such as the Ministry of the Environment. Geoparks could also gain by liaising with nonprofit conservation groups that have a proven track record on conservation. While geoconservation is a relatively new concept, conservation expertise from fields such as ecology and landscape conservation is highly relevant and applicable; experts working in those fields possess relevant knowledge and know-how for contributing to geoconservation. Indeed, as Matthews (2014) points out, conservation of geodiversity without regard to biodiversity and vice versa will not work out as in reality these facets are interrelated. Efforts should therefore be made to integrate insights from biological conservation, and especially habitat level conservation insights, with knowledge from earth science. This is especially important as we observed that large landscape units such as watersheds or coastal areas are currently under considerable stress.

Secondly, as Prosser (2013) points out in no uncertain terms, the fundamental part of geoconservation takes place through geoconservation audits. All geoparks, especially global geoparks, should provide such audits. Needless to say, rigorous assessment of current conditions of local geodiversity, current threats, and possible mitigation pathways needs to be implemented for such audits.

Thirdly, monitoring current conditions and ongoing change to geodiversity is fundamental for geoconservation. Geoparks must conduct routine monitoring based on expert advice. Existing legal schemes such as provisions for National Parks or Cultural Heritage Properties should be effectively used wherever possible, and if used properly, such provisions could strengthen geoconservation even without new laws.

Fourthly, in what follows from above, geoparks should come together to eventually formulate legally binding statutes that can open up new conservation and restoration pathways. This could be a medium to long-term goal, depending on the current level of preparedness and capacity of the geoparks.

Finally, as geoparks are part of a global initiative to celebrate the earth's heritage and geodiversity, Japanese geoparks could effectively liaise with their international counterparts to absorb lessons from good practices elsewhere. As indicated in the Introduction of this chapter, there are several instructive success stories of geconservation in Europe and these examples can be followed up or modified to suit the ground realities in Japan. Besides, professionals associated with global geoparks and geoconservation already possess relevant knowledge and experience that should be tapped into and utilized.

13.7 Conclusion

This chapter provided a description of the current challenges for geoconservation in Japan. The chapter described the findings from a survey that was conducted among global, national, and aspiring geoparks in Japan. Findings indicate that while there is a large number of national geoparks (and eight UNESCO accredited global geoparks) in Japan, geoconservation is not adequately grounded in their activities. The results identified numerous challenges such as lack of systematic threat evaluation, geoconservation audits, adequate planning structure, funding, and qualified conservation experts. While these pose significant challenges, there are some positive indications such as the existence of voluntary monitoring, protection of sites and features under existing laws, and signs that the UNESCO accreditation results in increased conservation awareness. It was also observed how geoconservation is a dynamic concept, but at the same time it is hard to be grounded at the local level. Current threats to geodiversity are manifold and result from lack of awareness in most cases. Whereas educating managers and local societies about conservation needs is important, it must be kept in mind that it is not enough to preserve specific landforms, landmarks, or sites and that conservation of the *diversity* of the natural environment is the ultimate task.

Acknowledgment We thank Yuichi Aoki of the Manifesto Research Institute of Waseda University for his assistance in statistical representation of findings.

References

Burek CV, Prosser CD (2008) The history of geoconservation. The Geological Society London, London

Chakraborty A, Mokudai K (2016) Emerging potential and challenges for geoconservation activities in Japan. ProGeo News 2016 (1):13–15

Chakraborty A, Cooper M, Chakraborty S (2015) Geosystems as a framework for geoconservation: the case of Japan's Izu Peninsula Geopark. Geoheritage 7(4):351–363

Christopherson RW (1992) Geosystems: an introduction to physical geography. Prentice Hall, New Jersey

Ellis N (2011) The geological conservation review(GCRI in Great Britain: rationale and methods). Proc Geol Assoc 122:353–362

Erikstad L (2013) Geoheritage and geodiversity management: the questions for tomorrow. Proc Geol Assoc 124:713–719

Graaf RD, Hooijmeijer F (eds) (2008) Urban water in Japan. CRC Press, Boca Raton, FL

Gracia Cortes A, Gallego E, Carcavilla L (2012) Spain. In: Wimbledon WAP, Smith-Meyer S (eds) Geoheritage in Europe and its conservation. ProGeo, Oslo, pp 334–343

Gray M (2004) Geodiversity: valuing and conserving abiotic nature. John Wiley, Chichester
Gray M (2013) Geodiversity: valuing and conserving abiotic nature. John Wiley, Chichester
Japanese Geoparks Network (English). http://geopark.jp/en/. Accessed 15 Mar 2017a
Japanese Geoparks Network. http://www.geopark.jp. Accessed 15 Mar 2017b
Kenny JA, Hayward BW (Eds) (1993) Inventory of important geological sites and landforms in the northern region. Geological Society of New Zealand Miscellaneous Publication, p 67
Knight C (2010) Natural environments, wildlife, and conservation in Japan Jpn Focus 8(4–2). http://apjjf.org/-Catherine-Knight/3292/article.html. Accessed 15 Mar 2017
Larwood JG, Badman T, McKeever PJ (2013) The progress and future of geoconservation at a global level. Proc Geol Assoc 124:720–730
Matthews TJ (2014) Integrating geoconservation and biodiversity conservation: theoretical foundations and conservation recommendations in a European Union context. Geoheritage 6 (1):57–70
Mokudai K (2014) Ugokituzukeru Daichi no Riyo to Taiji (Using and facing the reality of the continuously moving Earth). In: Miyamoto S, Nonaka K (eds) Shizen to Ningen no Kankyo Shi (Environmental history of nature and human society). Kaiseisha, Otsu pp 195–212 (In Japanese)
Nakada S (2017) Volcanic Archipelago: volcanism as a geoheritage characteristic of Japan. In: Chakraborty A et al (eds) The natural heritage of Japan: geological, geomorphological, and ecological aspects. Springer, Berlin
Prosser CD (2013) Our rich and varied geoconservation portfolio: the foundation for the future. Proc Geol Assoc 124:568–580
Scottish Geodiversity Forum (nd) https://scottishgeodiversityforum.org Accessed 15 Mar 2017
UNESCO (nd) Part B: UNESCO Global Geoparks. http://www.unesco.org/new/fileadmin/MULTIMEDIA/HQ/SC/pdf/IGGP_UGG_Statutes_Guidelines_EN.pdf Accessed 15 Mar 2017
Uzawa H, Okuma T (Eds) (2010) Shakaiteki Kiyotsu Shihon toshite no Kawa (Rivers as social common capital). University of Tokyo Press, Tokyo (In Japanese)
Wakita K (2017) Geology of the Japanese Islands: an outline. In: Chakraborty A et al (eds) The natural heritage of Japan: geological, geomorphological, and ecological aspects. Springer, Berlin
Watanabe M (2017) The origin and development of geoparks in Japan: reflections from a personal perspective. In: Chakraborty A et al (eds) The natural heritage of Japan: geological, geomorphological, and ecological aspects. Springer, Berlin
Whiteley MJ, Browne MAE (2013) Local geoconservation groups: past achievements and future challenges. Proc Geol Assoc 124:674–680
Worton GJ, Gillard R (2013) Local communities and young people: the future of geoconservation. Proc Geol Assoc 124:681–690
Yoshikawa T (1984) Geomorphology of tectonically active and intensely denuded regions. Geogr Rev Jpn 57A:691–702 (in Japanese with English abstract)

Part III

National Park Sites

The Lake Akan Area: A Future Geopark? 14

Malcolm Cooper

Abstract

Akan National Park is found on the Island of Hokkaido. It was designated in 1934 as one of the second batch of properties to be declared national parks in Japan. The natural heritage of the area includes a group of volcanoes, Me-akandake, O-akandake, and Akan-Fuji, and three lakes of volcanic origin, Akan-ko, Kussharo-ko, and Mashu-ko. The geological structure consists of a large caldera that originated in the early Pleistocene and a group of younger partly Holocene, andesitic, and dacitic cones. The highest point of the complex is the Me-akandake (1499 m) stratovolcano. It consists of nine overlapping cones, with three summit craters, and has erupted least 17 times since the beginning of the nineteenth century. The national park based on this natural heritage is a popular visitor destination, offering sought-after views and bathing opportunities all year round and skiing and other snow sports in the winter. Lake Akan-ko is also known for its "marimo"—a species of algae that forms large green balls when mature—and a natural phenomenon that in Japan is unique to this lake. The social capital of the area also includes traditional Ainu settlements (an early Japanese indigenous ethnic group now mainly confined to Hokkaido). This is an important feature given that cultural geotourism is a global phenomenon. Many indigenous communities offer their unique interpretations of natural and other forms of heritage, including food, around the world, and the implications of such an interest by tourists for the Ainu community around the lake are considerable.

Keywords

Marimo • Ainu • Cultural interpretation • The Mt. Akan volcano group • Geoparks

14.1 Introduction: The Dynamic Volcanic Environment of Akan and its use in Tourism

Akan National Park is located in the north-eastern part of Hokkaido, near to the Pacific coast. It was designated a national park in 1934, and has a total ground area of 904.8 km^2 (90,481 hectares). The natural heritage of the area is dominated by a group of volcanoes, including the Me-akandake complex (with Akan-Fuji), O-akandake, and Atosa Nupuri, and three volcanic lakes: Akan-ko; Kussharo-ko, and Mashu-ko. The total package is known as the Akan volcanic complex and is an active volcanic environment (see Fig. 14.1 and Table 14.1). The geological structure of this area consists of a large caldera that began activity in the Pleistocene and a group of younger partly Holocene andesitic and dacitic cones. The highest point of the complex is the Me-akandake (1499 m) stratovolcano. It consists of nine overlapping cones, with three summit craters, and has erupted least 17 times since the beginning of the nineteenth century (recent eruptions include those in 2015 at Me-akandake and in 2008 at O-akandake (Cooper 2014).

M. Cooper (✉)
Ritsumeikan Asia Pacific University, Beppu, Japan
e-mail: cooperm@apu.ac.jp

A. Chakraborty et al. (eds.), *Natural Heritage of Japan*, Geoheritage, Geoparks and Geotourism,
DOI 10.1007/978-3-319-61896-8_14

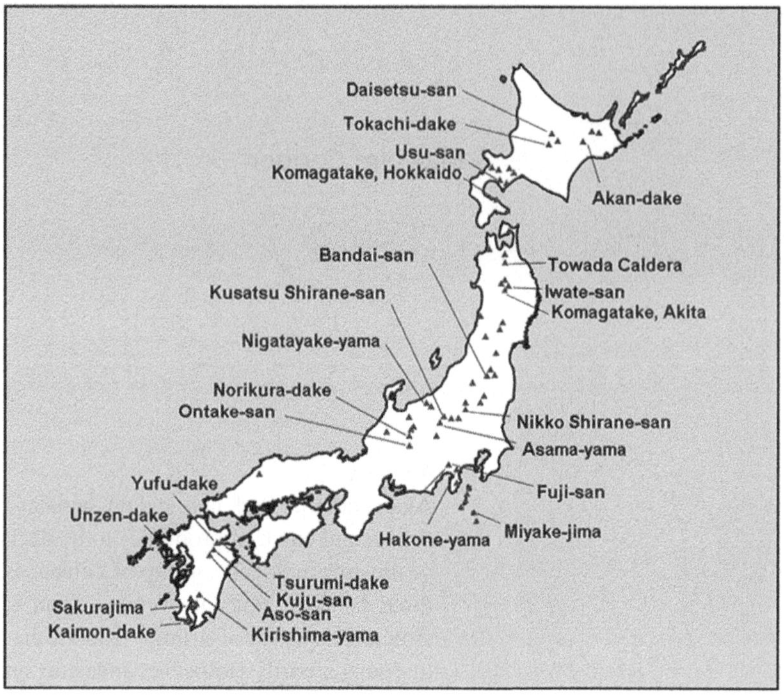

Fig 14.1 Distribution of the major active volcanoes of Japan. Source: Geological Survey of Japan/AIST as modified by Cooper (2014)

Table 14.1 Akan Volcano eruptions since 1800

Date	Nature
2015	Ash and "bombs"
2008	Ash and lava (small scale)
2006	Ash
1988, 1996, 1998, 2008	Phreatic
1964–1966	Phreatic explosion
1962	Phreatic explosion
1954–1960	Phreatic explosion
1951–1952	Unknown
1927	Unknown
1808	Unknown
1800	Phreatic explosion

Source: http://www.volcanodiscovery.com/akan.html; www.jma.go.jp/jma/indexe.html

Research has shown that there have also been at least four major magmatic eruptions with pyroclastic flows from this mountain during the past 10,000 years (Cooper 2014). Added to this should be the Holocene eruptions of O-akandake that extensively modified the environs of Lake Akan (Fig. 14.2).

Hiking is a favorite activity in the park. The two Me-akandake access trails are found on the western side of that mountain. The hiking groups generally depart from one and return via the other, but choice of direction is free. O-akandake has a trail that starts from the eastern end of Lake Akan (Japan-guide.com 2016). Hiking to the top of each mountain takes about 6 h, so they are not difficult to climb, and the trails are certainly passable June to October. In other months, hiking for inexperienced climbers is usually not possible, as the trails are likely to be buried under snow. The park is also a very popular destination for other forms of tourism: offering panoramic views and hot springs for bathing all year round and skiing and other snow sports in the winter. Akan-ko is known for its sublime beauty but also for its "marimo"—a type of algae that forms large green balls when fully grown, and a natural phenomenon that in Japan is unique to this lake. Left undisturbed for hundreds of years, Lake Akan's marimo balls grow to about 20 cm in diameter (Irimoto 2004). The algae have been designated a National Special Natural Monument. Of the other lakes, Mashu-ko is reputed to have the highest level of clarity in the world, and there is also an old-growth forest of spruce and white fir trees in the vicinity, along with the volcanoes.

14.2 The Influence of Community Natural Heritage

Of importance to our recognition of the importance of community natural heritage is how the communities themselves interpret their structure, meaning, and value. The Lake Akan area is best characterized as a form of mixed natural and social capital because, almost uniquely in Japan, this area includes traditional Ainu communities (a Japanese indigenous ethnic group now mainly confined to Hokkaido) alongside those of the mainstream population (Chang et al. 2011). This is an important feature given that cultural tourism has become of worldwide interest as indigenous communities offer interpretations of natural heritage, through their unique

Fig 14.2 Lake Akan and environs. Source: Photo courtesy of the author

cultural performances, and other natural heritage commodities such as food, to add to the appeal of the wider communities (Cheung 2005). The implications of such an interest by tourists on the Ainu community around Lake Akan have been explored through the application of a stage model of creative cultural destruction. Chang et al. (2011) note that, with the colonization of Hokkaido after the Meiji era (1868–1912), the indigenous Ainu way of life has been negatively affected by the national assimilation policy. However, the past 60 years has seen tourism to Ainu settlements grow, and Ainu have begun to host tourists in their traditional costumes at the various tourist destinations in Hokkaido. Lake Akan is no exception to this trend.

By collecting data on tourist arrivals, on investment by entrepreneurs in facilities, and on local community attitudes, it has been shown that the Lake Akan Ainu are in the model's second stage, advanced commodification, of the process of the creative destruction of an indigenous community (Cheung 2005). This model of change postulates that an indigenous community will go through various responses to the possibilities of economic development through trade based on their culture, before perhaps disappearing as a sustainable traditional culture in favor of complete assimilation into the mainstream. The resulting cultural tourism and hospitality overlay on the area's natural heritage is attractive to visitors and shows how interest in natural heritage can be aligned with cultural expressions of their use in the local area. This is a useful finding for other communities proposing to base their tourism development on cultural interpretations of natural heritage. The Ainu variant includes performances at their "kotan" (village) and control of various leisure activities such as canoeing, mountain biking, and camping. These natural and cultural attractions, together with the hot spring baths on the lakeside, may qualify the Akan National Park area for inclusion at least in the Japanese National Geopark Network, if not the global one.

The village of Akan-kohan is the only settlement close to Lake Akan. It offers several high-quality and expensive ryokan-style accommodation facilities, some of which have bathing facilities allowing direct views of the lake, and/or extends the use of their onsen facilities to casual bathers for a fee during the day. At the town's eastern end, it is possible to find hot mud pools locally known as *bokke* (Fig. 14.3). Access is gained from the *Akankohan Eco Museum Center,* where material on the National Park, its natural geoheritage, and its wildlife and local culture is available. In relation to the social heritage of the area, Ainu-Kotan is a small Ainu precinct in the resort village (some 200 inhabitants); it is primarily a street lined by shops selling Ainu handicrafts (japan-guide.com 2016). In this street, there is a small museum of traditional Ainu crafts, clothes, and how daily life was carried on in earlier times. Traditional cultural performances are also held at various times in a hall next

Fig. 14.3 Bokke mud pools. Source: Photo courtesy of the author

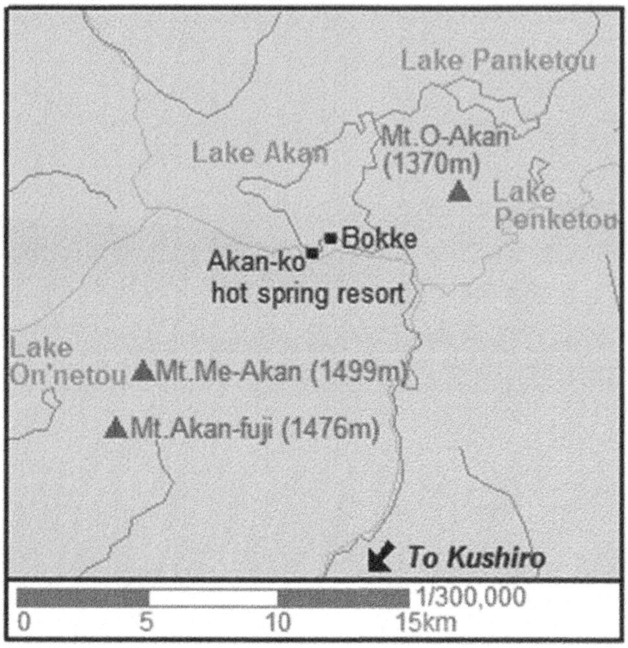

Fig 14.4 The Lake Akan area in detail. Source: Cooper (2014)

to the museum. Finally, the town offers short boat trips around the lake, which include a stop at the *Marimo Exhibition Center*, and access to the more remote parts of the lake.

14.3 Lake Akan: A National Treasure

Akan-ko is a lake located in the southwest of the Akan National Park, between the two volcanic masses of Me-akandake and O-akandake. Lake Mashu and the Mashudake complex lies at the opposite end of the Park, north-northwest of Kushiro City, Hokkaido. The area has been designated the Akan National Park and listed as a registered wetland under the Ramsar Agreement (2005).

Originally, Akan-ko was a much bigger caldera lake that formed after a very long caldera-forming sequence of particularly high explosivity volcanism spanning 1 million years (Moreno et al. 2012). The major eruptive phases are sequenced from AK 17 to AK 1 dating from 1.46 Ma to 0.21 Ma (Ibid), but the defining shape of the Akan caldera was likely formed out of AK 2 (nearly 175,000 years ago) and AK 1 eruptions (nearly 158,000 years ago). Quaternary volcanism such as that of the Me-akan volcano (1499 m), located southwest of the current lake (Fig. 14.4), has since dominated this volcanic landscape. Around 10,000 years ago, the volcano O-akandake (1370 m), located east of the current lake, erupted, fragmenting the old lake. The result was the present set of major water bodies in the National Park: Lake Akan, Lake Panketou, Lake Penketou, and a few other small lakes (Fig 14.4).

Apart from its scenic quality and pure water, Lake Akan is famous for the size of the algae colonies found there. *Aegagropila linnaei*, known locally as Marimo in Japanese, and as a Cladophora or Moss Ball in English, is a species of Chlorophyta found mainly outside Japan, in lakes in northern Europe (Wakana 1992). The marimo is, however, a rare form of this species where the algae grow into large green balls (Fig. 14.5). Other than in Japan, this type is known only in Iceland, Scotland, Estonia, and Australia (Einarsson et al. 2004).

Marimo were first discovered in the 1820s in Lake Zell, Austria (Wakana 1992; Yoshida et al. 1994). Their classification in the genus *Aegagropila* was decided on in 1843, with the name *A. linnaei* adopted as the type species based on their physical configuration in the shape of a ball. The algae were named "marimo" by Japanese botanist Tatsuhiko Kawakami in 1898 (Hanyuda et al. 2002). A *Mari* is a ball, and *Mo* is a generic Japanese term for plants that grow in water. The Ainu names are *torasampe* (lake goblin) and *tokarip* (lake roller) (Irimoto 2004). Marimo balls are occasionally for sale to private collectors under the name "Japanese moss balls," even though they are not moss (Yoshida et al. 1994). There are three growth forms of this species of algae: one grows on rocks in a lake, usually on the shady side of this environment. Another is *free-floating filaments* in the water mainly, but small patches of unattached algae can form a dense covering on the bottom of a lake. The third form is the moss ball, where the algae can grow into 20–30 cm round collections that have no central structure (Hanyuda et al. 2002).

For their survival, the marimo colonies in Lake Akan depend on their successful adaptation to low light conditions, to the wind-induced currents found on the lake, to lake morphology, to the bottom substrate, and to the sedimentation and other processes at work in the wider lake environment (Hanyuda et al. 2002). Their annual growth rate is about 5 mm (Wikipedia 2016), and they can

Fig. 14.5 Marimo Algae. Source: Photo courtesy of the author

grow to between 20 and 30 cm in diameter (Fig 14.5). Their characteristic shape is maintained by the slight wave action found in Lake Akan, because this is enough to turn the balls over from time to time. This action also cleans them. At Lake Akan, effort is devoted to the preservation of this form of algae, including an annual 3-day festival in which Ainu culture plays an important part in illustrating the value of the marimo to the local community and visitors. Because of their interesting appearance, these algae also serve as a medium for the education of locals and tourists alike in relation to the environment in the museum and in schools.

In Japan, the marimo have been protected since 1920 and are now defined as a *natural treasure* (Irimoto 2004). The small versions sold as *omiyage* (souvenirs) are made from the free-floating filaments found in the Lake. Because of domestic tourism and the requirement for omiyage associated with this, awareness of the marimo is quite high across Japan. In line with the norms of "authentic" Japanese tourism tradition, a stuffed toy character known as *Marimokkori* commemorates the lake balls. This is yet another illustration of the crossover between items of natural heritage and cultural values that persists even today and does not require the intervention of mythology or religion to be sustained.

14.4 Other Sites of Interest in the Akan National Park

O-akandake (1371 m) is in the middle of the park, adjacent to the next geological complex of importance in Hokkaido, the Chishima Volcano Belt. It first appeared about 8000 BCE in the interior of the then much larger Akan Caldera. The striking conical shape of O-akandake is known in the Ainu language as Pinneshiri (Male Mountain), an indication of how these geosites are often seen in mythology as creation devices. The mountain harbors a mixed conifer and broadleaved forest to about the 1000 m level and the rest is open grassland and dwarf pines interspersed with large and small deposits of lava (Cooper 2014). Above the seventh station on the major trails on the mountain, hikers can see very good views of the natural heritage of Lake Akan and Me-akandake on clear days.

Me-akandake is found adjacent to and southwest of Lake Akan and rises to 1499 m. Volcanism began some 15,000 to 20,000 years ago, and, in common with many Japanese volcanoes, Me-akandake comprises multiple cones, including Minamidake, Higashidake, and Tsurugidake, and vents, of which the most important ones are Nakamachineshiri and Ponmachineshiri (the site of the most recent activities). Corresponding to the male designation of O-akandake in the Ainu language, Machineshiri is the "Female Mountain." Me-akandake's main summit, Ponmachineshiri, formed 3000–7000 years ago, while Akan-Fuji to the south side of Me-akandake (1476 m) first erupted about 2000 years before the present. More recently, phreatic eruptions occurred at this site between 1955 and 1959 and 2006 and 2008.

The word "Bokke" translates as "boil" in Ainu and describes the activity of the mud volcanoes found near Lake Akan (Cooper 2014). The Bokke area of the lake is located about 500 m from the Akan Lakeside Eco-Museum Center. The base geology of this area features numerous hydrothermal discharge spots from which gas and water are continuously discharging. A mixed forest containing a diverse ecosystem is found around these hydrothermal outlets.

Yu-no-Taki is a further attraction situated about 1.5 km south of Akankohan. Yu-no-Taki is made up of two waterfalls, with the left carrying hot water from a vertical height of 30 m (at a temperature of 43 °C) (Cooper 2014), and was designated as a natural monument of Japan in September 2000. The waterfalls have co-located tourist facilities such as toilets. Yu-no-Taki was used as an open-air bath in the past, but using the waters in this way was stopped when it was discovered that the site is within a naturally formed deposit of manganese (Cooper 2014). The black mud on the surface of the site is manganese oxide.

Heritage in a Nutshell: Akan National Park (Lake Akan and its Surrounding Landscapes)
Type of Heritage: Composite: Geological-Geomorphological-Ecological Heritage Complex involving active volcanoes, older caldera formation, lakes, marimo colonies (the only such natural colony in Japan), as well as cultural attractions (Ainu cultural artifacts, representations, folklore). A part of the area is also designated as a Ramsar wetland (since 2005).

Akan National Park (nearly 904 km^2) was designated in 1934. Popular to hikers throughout the year, the park has two main types of natural heritage attractions, i.e., Lake Akan (with its marimo colony) and adjacent lakes and the Me-akan-O-Akan volcanic complexes. Both Me-Akan and O-Akan volcanoes have accessible and popular trails; the Me-Akan trail takes hikers past fumaroles and other signature sites of active volcanism. The area is also popular for winter sports and hot-spring bathing. Lake Akan is famous for its panoramic vista and as the habitat of *Aegagropila linnaei* or marimo (algae) balls that attain strikingly beautiful rounded shapes. Lake Akan is located on a larger, older caldera formation that began with high explosivity volcanism in the Pleistocene. The Akankohan Eco Museum Center offers useful information about the area's natural heritage.

In addition, visitors can enjoy the cultural landscape of the Ainu community; Ainu cultural performances can be experienced at a theme village (Ainu Kotan). The area is a well-known hot-spring hub, and excellent lodging facilities are available.

Access: Accessible from Kushiro, Memanbetsu, and Tokachi domestic airports. Shuttle bus from Kushiro Airport to Lake Akan takes 90 min.

Useful websites: Lake Akan Tourist Information: http://www.lake-akan.com/en/

14.5 Conclusions

The first National Parks Law in Japan was enacted in 1931. In the middle of 1934, the first areas to be officially designated as national parks in Japan were listed, while Akan was listed in December 1934. The Akan park is 90,481 ha in extent, and its natural heritage highlights include the Lakes Akan, Mashu, and Kussharo, the volcanoes Me-akandake, O-akandake, and Atosa Nupuri, mud pools, and onsen. In addition, it boasts some of the largest specimens of the algae known as Marimo or moss balls, which grow to a maximum size of around 30 cm.

The geological heritage of this area consists of a large formerly water-filled caldera and a group of younger quaternary volcanoes, along with smaller caldera lakes. The highest point of the complex and a very active area is the Me-akan (1499 m) stratovolcano. The national park based on this natural heritage is a popular visitor destination, offering sought-after views and bathing opportunities all year round and skiing and other snow sports in the winter.

The social heritage of the area includes traditional Ainu settlements (an early Japanese indigenous ethnic group now mainly confined to Hokkaido). This is an important feature given that cultural tourism is a global phenomenon. Many indigenous communities offer their unique interpretations of natural and other forms of heritage, including food, around the world, and the implications of such an interest by tourists for the Ainu community around the lake are considerable. This community already engages in tourism that enhances the natural heritage of the area through its cultural traditions.

Acknowledgement The author thanks Dr. Wataru Hirose of the Environmental and Geological Research Department, Geological Survey of Hokkaido, for his welcome contributions to the geological commentary in this chapter.

References

Chang J, WY S, Chang CC (2011) The creative destruction of the Ainu Community in Hokkaido, Japan. Asia Pac J Tour Res 16 (5):505–516

Cheung S (2005) Rethinking Ainu Heritage: a case study of an Ainu settlement in Hokkaido, Japan. Int J Herit Stud 11(3):197–210

Cooper MJ (2014) Volcanic national parks in Japan. In: Erfurt-Cooper PJ (ed) Volcanic tourist destinations. Springer, Heidelberg, pp 231–246

Einarsson A, Stefánsdóttir G, Jóhannesson H, Ólafsson JS, Gíslason GM, Wakana I, Gudbergsson G, Gardarsson A (2004) The ecology of Lake Myvatn and the River Laxá: variation in space and time. Aquat Ecol 38:317–348

Hanyuda T, Wakana I, Arai S, Miyaji K, Watano Y, Ueda K (2002) Phylogenetic relationships within Cladophorales (Ulvophyceae, Chlorophyta) inferred from 18S rRNA gene sequences, with special reference to Aegagropila Linnaei 1. J Phycol 38(3):564–571

Irimoto T (2004) Creation of the Marimo festival: Ainu identity and ethnic symbiosis. Senri Ethnol Stud 66:11–38

Japan-guide.com (2016) Lake Akan. http://www.japan-guide.com/e/e6801.html? Accessed 20 Nov 2016

Moreno T, Willis S, Kojima T, Gibbons W (2012) The geology of Japan. The Geological Society, London

Wakana I (1992) A bibliography relating to “Marimo” and their habitats. Marimo Res 1:1–12

Wikipedia (2016) Marimo. https://en.wikipedia.org/wiki/Marimo? Accessed 20 Nov 2016

Yoshida T, Nagao M, Wakana I, Yokohama Y (1994) Photosynthetic and respiratory property in the large size spherical aggregations of “Marimo”. Marimo Res 3:1–6

Oze Wetland: The Birthplace of the Nature Conservation Movement in Japan

15

Kiyotatsu Yamamoto

Abstract

This chapter describes how the modern nature conservation movement in Japan grew out of opposition to dams and other development schemes in the Oze Wetland in Honshu. The Oze wetland is a rare highland marsh with a distinctive landscape and biodiversity. The region was virtually unknown until the industrial development of the late nineteenth century, but thereafter its environment was rapidly placed under threat from several development schemes, the most contentious of which was a dam construction proposal. However, Oze became a rare example where bureaucrats, nature conservationists, and scholars came together to form a successful opposition movement that succeeded in saving the wetland; this movement was subsequently broadened at the national level. However, the rise in the popularity of Oze also meant a rise in visitor impact on its natural environment. The chapter outlines the main events in the course of the development of the Oze conservation and management program and identifies the problems that make active conservation of this precious environment a continual need.

Keywords

Wetland environment • Nature conservation movement • Visitor pressure • Oze Wetland

15.1 The Nature of Oze

Due to the topography of the Japanese Islands, low-lying lands are rare, and for this reason, marshes or wetlands were subjected to reclamation and cultivation from ancient times. Among the remaining wetlands, Oze (Oze-ga-hara) is among the largest, apart from those found in Hokkaido Island (Fig. 15.1). A list of 20 large wetlands of Japan reveals that most of the large wetland areas are located in Hokkaido, with Kushiro Wetland (227 km^2) being the largest of all, and the only entries from outside Hokkaido are Watarase (20 km^2) and Oze. During the Meiji (1868–1912) and Taisho (1912–1926) Periods, the total area covered by wetlands in Japan was 2111 km^2, but that area dwindled to only 821 km^2 in 1999 (Geospatial Information Authority of Japan 2015). Due to this rapid decline in the wetland environment, Japan's wetlands have a special value for conservation of the archipelago's biodiversity and geomorphological diversity.

Within the area known as "Oze," the "Oze-ga-hara" (literally Oze highland marsh) is the actual wetland. It is the largest high-altitude wetland in the Honshu Island, surrounded by the two peaks of 2356 m asl Hiuchigatake (locally perceived as the "father mountain" due to its imposing appearance) and the 2228 m asl Shibutsusan (locally perceived as the "mother mountain" due to its gentle appearance). The mean elevation of the wetland is 1400 m asl; the wetland has a total area of 7.6 km^2 with a east-west spread of 6 km and a north-south spread of 1 km (Fukada 1982). A large number of endangered

Electronic supplementary material: The online version of this chapter (doi:10.1007/978-3-319-61896-8_15) contains supplementary material, which is available to authorized users.

K. Yamamoto (✉)
Department of Environmental Sciences for Sustainability, Faculty of Agriculture, Iwate University, Iwate, Japan
e-mail: kiyo@iwate-u.ac.jp

A. Chakraborty et al. (eds.), *Natural Heritage of Japan*, Geoheritage, Geoparks and Geotourism,
DOI 10.1007/978-3-319-61896-8_15

Fig. 15.1 High-altitude wetland in Oze

and rare vegetation species inhabit this area: these include the *Nuphar pumilum* var. ozeense (Oze-kouhone), *Japonolirion osense* Nakai (indigenous to the Shibutsusan mountain), and popular flowering plants such as *Lysichiton camtschatcense* Schott (Asian skunk cabbage, a plant that is almost symbolic of Oze) that blooms in June, *Hemerocallis middendorffii* var. esculenta (Daylily) that blooms in early July, and several broad-leaved tree species known for their fall colors (Natural Parks Foundation 2009).

A geographical characteristic of Oze is that the wetland is located in a comparatively high-altitude area of the Honshu island of Japan and supports a large number of native species. The easy accessibility of Oze from Tokyo means it is visited by large numbers of tourists throughout the year. This in turn creates overuse pressure on a landscape that is an important habitat for rare species. The most notable threat to the wetland, apart from tourism-related overuse (Nakajima 1998), are damage to plants due to trampling, sewage problems, invasive species, and overgrazing by Shika deer (*Cervus nippon*) (Natural Parks Foundation 2009).

15.2 A Brief History of the Nature Conservation Movement

The Oze National Park itself is relatively new (Table 1); it came into existence in 2007 as the 29th national park of Japan, when the area was separated from the Nikko National Park. The hallmarks of this national park include landscapes of exceptional beauty and high biodiversity.

However, Oze had played a major role in shaping the nature conservation movement in Japan for a long time prior to its listing. It is also referred to as the "birthplace of nature conservation movement in Japan" by many. This is because of the activities of the "Oze Hozon Kisei Domei" (Association for the Preservation of Oze) that began in 1949 as the protest movement against a planned dam at the site; this organization was the predecessor of the Nature Conservation Society of Japan (NACS-J) which currently spearheads nature conservation movements at the national level (the NACS-J was born in 1951, after the Kisei Domei was dissolved that year) (NACS-J

1985). The main objective of the Kisei Domei was to preserve the natural environment of Oze from the threat of a hydroelectricity project. The issue of energy development and the conservation of a landscape of outstanding beauty became a highly contentious one, and along with similar problems surrounding the development schemes of the Kurobe and Kitakami River Valleys, it reverberated in contemporary society and affected the politics of that time (Tamura 1954). Thus, the movement has a relatively long legacy as a nature conservation scheme in Japan.

However, the two national park visitor centers in the area currently provide little guidance on the rich history and value of the place as a nature conservation hub of modern Japan. During the skunk-cabbage flowering season, a survey was carried out by my laboratory's survey team among 288 respondents who visited the area; the survey revealed that while 59.7% knew of Oze's role as the birthplace of nature conservation, 79.8% responded that they would like to see that aspect more emphasized in the guidance. This means that while the nature conservation history of Japan is relatively well known, there is a perception that this history is not shared well among visitors. While it is difficult to pinpoint where exactly the nature conservation movement of modern Japan began, the following historical events that signify an early start to nature conservation are worth mentioning in regard to Oze.

According to Hatakeyama (2001), the movement to preserve the Oze wetland from being submerged in a dam reservoir became a key connecting factor to the nature conservation movements of post-WWII Japan. Hatakeyama points out that the nature conservation movement took shape as a protest movement against the comprehensive hydroelectricity development plan in the post WWII period and that the formation of the Kisei Domei as an organization devoted to this cause was a key feature in the history of nature conservation movements. Murakushi (2005), on the other hand, points out the key roles played by people such as Chozo Hirano (often credited as one of the pioneers of nature conservation in Meiji Era Japan); several scholars, educators, mountaineers, and publishers who followed Hirano; enlightened bureaucrats from the internal affairs, agriculture and forestry, and culture ministries, as well as the contributions from the national park-related organizations; and the committee for ascribing historical and natural landmarks as key factors behind the success of the nature conservation movement. According to Numata (1994), the fact that the Kisei Domei was a private organization was key for nature conservation at that time. The famous mountaineer and botanist of the generation, Hisayoshi Takeda, mentioned that the Japanese Alpine Club had also played a pivotal role in the formation of the conservation agenda in Oze (Ishikawa 2001). However, there is one point that ought to be mentioned with regard to Numata (1994)'s claim: the Kisei Domei was not a completely private organization. The Contemporary Culture and Health and Welfare Ministries had two and three representatives, respectively (NACS-J 2002). Because the Kisei Domei was related to the birth of the NACS-J which became the main nature conservation association in postwar Japan, Oze can be considered as the "birthplace" of such activities (Hatakeyama 2001; Numata 1994; Ishikawa 2001). Based on this logic, it is acceptable that we recognize Oze as a pioneering place that shaped the post-WWII nature conservation movement and its discourse.

15.3 Oze's Legacy in Nature Conservation

Although Oze is a highland marsh and the Alpine Club of Japan played an important role in the conservation of this area, Oze does not have a particularly long history of association with mountaineers. Furthermore, the region was made accessible only after the famous alpinist and nature conservationist Chozo Hirano opened a mountain hut in 1889. Oze does not have the long history of pilgrimage visits out of faith either, which puts the area in contrast with other popular destinations. It is also thought that the area was virtually unknown before it was featured in the magazine "Taiyo" (Sun) in 1895 (Kikuchi and Sudou 1991). Oze became widely known only when the hydroelectricity development plan was formulated in 1903 and the struggle between the developers and the nature conservation activists was reported in the contemporary media. On the other hand, Oze enjoyed a swift transformation as a national park; a petition for designating the area (the Nikko area) was presented to the imperial parliament in 1912 (Bansho 2013). This indicates that the conservation values of Oze and Nikko were recognized at that point. The area was eventually designated as a national park in 1934, following the Seto Naikai, Unzen, and Kirishima National Parks and along with Akan, Daisetsuzan, Chubu Sangaku, and Aso national parks. In 1949, the NHK aired a program on the area that led to a rapid rise in the popularity of Oze. The rise in the popularity was reflected in the increase of the visitor numbers, and the Ministry of Health and Welfare that was the principal organ behind national parks' designation also recognized the value of Oze along with other major tourist destinations such as Nikko, Fuji-Hakone, and Seto Naikai (Inland Sea). This combination of various factors paved the way for the recognition of Oze's landscape within the bureaucracy.

Thus, it is important that we emphasize the events that immediately preceded or followed the 1949 announcement of the hydroelectric dam project. In February 1947, the Commerce and Industry Ministry passed a unilateral order of building the hydroelectric dam without referring to the opinions of the relevant ministerial organs such as the

Fig. 15.2 Oze before the establishment of walkways

Ministries of Interior, Education, and Health and Welfare. This development was immediately met with opposition; in a meeting on 4 July 1949, opposing voices centered around the opinion of Masayoshi Takeda were raised at the Education Ministry. Tsuyoshi Tamura, the most influential figure behind the designation of Japan's national parks, also weighed in with his negative observations at the Health and Welfare Ministry. On 23 July 1949, a verbal pledge was secured from the developers, who agreed not to disturb the natural environment of Oze in return for access to the water resources from the marsh. However, on 19 February 1949, a proposal to build an 80 m tall dam that would transform the entire wetland into a big reservoir came to light; this implied that the verbal pledge was never seriously considered as binding by the developers. The education and health and welfare ministries joined hands to stop the project and started raising awareness about Oze through a pamphlet entitled "On the Scientific Value of the Oze Wetland." On the other hand, the developers invited experts from the USA to implement a feasibility study regarding hydroelectricity generation at Oze. The Kisei Domei was born at this juncture on 27 October 1949. The list of the 27 initial members included scientists, agricultural experts, mountaineers, artists, photographers, and the Chozo Hut (set up by Chozo Hirano) managers. The organization subsequently grew and played its role of opposing the development project by raising signatures. In 1951, after its immediate goal was achieved; the Kisei Domei was dissolved and the NACS-J was born with a wider mandate of nature conservation across Japan.

In the following years, the number of visitors to Nikko National Park increased steadily (Figs. 15.2 and 15.3). In 1952, the Yamaguchi Forestry Office undertook a project to construct wooden walkways (Fig. 15.3). However, some problems remained, such as the contrast between the approaches of the Fukushima and Gunma Prefectures. Whereas in the Fukushima Prefecture there was an early effort to develop trails for safeguarding plant colonies with Forestry Agency funding (due to the designation of the land as a national property), such efforts did not materialize in the Gunma Prefecture (where the land belonged to private operatives), leading to problems of unregulated trampling of plants by visitors (The Asahi Shimbun 1957). This indicates that at that initial stage planning and management styles of the national and prefectural organs were poorly coordinated and the management structure was still in its infancy. In 1953, the first ranger (1 out of 6 initial "rangers" assigned for national park management that year) was stationed at Nikko Yumoto. In that same year, Oze was assigned as a specially protected area within the national park. Nature conservation and management efforts were further strengthened as Oze became a natural monument in 1956 and a special natural monument

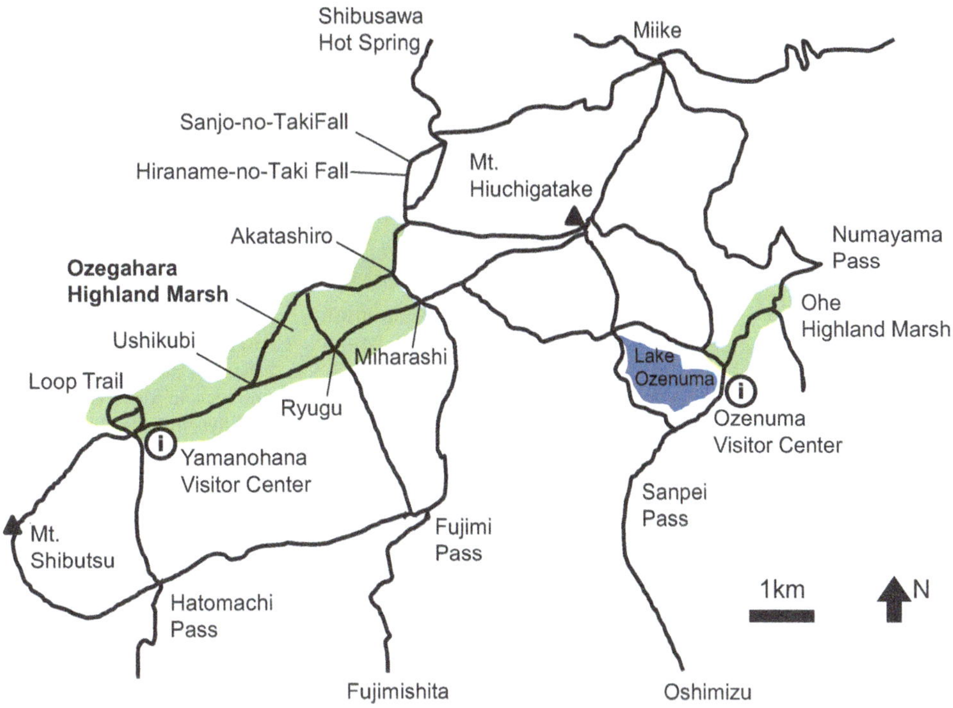

Fig. 15.3 Trail map of Oze marsh

(raising its value) in 1960. However, during this period so many visitors flocked to Oze that the modest mountain huts could not absorb the pressure, and serious problems due to trampling and littering of waste arose. This led to a clean-up drive, which was another pioneering activity as far as managing national parks is concerned.

From 1960, most visitors traveled to Oze by car, and the Oze-Oku Tadami Tourism Association was created in 1965 jointly by Fukushima, Gunma, and Niigata Prefectures. Although this was an effort to promote tourism and local development, the environmental problems at Oze were also highlighted in this process. In 1965, the Committee for the Protection Cultural Properties began considering whether to restrict some access to the preserved area (The Asahi Shimbun 1965). The Health and Welfare Ministry, which was in charge of national park management, added that it would not allow cable car lines to be constructed at Oze, and a statement from that ministry proposed doubling the extent of walkways to protect plant colonies (The Asahi Shimbun 1966). In 1966, a committee for conserving the wetland came into existence at the Gunma Prefecture; this committee highlighted the denudation of vegetation in the Ayamedaira area and began efforts for landscape restoration.

In 1970, a second threat appeared: in response to the rapidly growing tourist numbers, a plan to construct a roadway at the outskirts of the special protection zone was mooted. In 1971, Director General of Environment Agency Buichi Oishi conducted a spot survey and demanded either the plan be scrapped or the road be redirected to a different location (Hirano 1972). This was reported in the media and soon a momentum to protect the natural environment had built up across the country. Finally, in August 1971 the road construction plan was officially scrapped after a direct petition was submitted to Oishi by the then Chozo Mountain Hut owner and nature conservationist Chosei Hirano, and a signature campaign against the project took place. In 1972, a program to educate visitors to carry their waste back was launched, and in the same year, Gunma Prefecture formulated the Oze Charter, a pledge to protect the area (Oze Preservation Foundation 2002). In 1974, car access restriction was brought into effect in the Hatomachi and Numayama Pass areas (Furuya et al. 2001); this was another pioneering initiative at the national level. In this way, several conservation initiatives were launched, but in effect, the waste problem could not be remedied swiftly and several years were required to gradually phase out the litter bins. From the 1980s, water quality standards were determined, and in the following years, restriction on using soap and shampoo was introduced in the mountain huts (The Asahi Shimbun 1983).

In 1988, efforts to address the problem of visitor overuse began, and demands were soon made for swift implementation of user restriction and payment of access fees. From this time, local administrative units also started collaborating on management, and a collaborative management framework came into existence. In 1988, the "Committee for Conservation of the Oze Area of the Nikko National Park" was formed with participants from the Bureau of the Environment and the Fukushima, Gunma, and Niigata Prefectures. In 1989, sensors to count visitors on the trail were established. However, the proposed fee collection scheme could not be implemented due to opposition from the Japan Workers' Alpine Federation and local administrative bodies. In 1992, an "Oze Summit" was held in the Ozenuma Mountain Hut, and in 1995, the Oze Preservation Foundation was born (Oze Preservation Foundation 2002). In 1996, a total of 647,523 visits were recorded in Oze; this was a record number (MoE 2016). From this point, the Oze Preservation Foundation began surveys to understand visitor impacts and control the same by dispersing visitors through schemes such as differentiating weekend and weekday lodging costs, and implementation of "no bathing" days, in coordination with the mountain huts. In the fall of 2012, a survey was carried out (Yamamoto and Kitabatake 2012) on visitor perceptions in the aftermath of the Fukushima Daiichi nuclear disaster: this survey found that 31 of the 369 surveyed had concerns over local radioactive contamination. Some typical answers were: "I searched for the radioactivity levels beforehand," "I changed my access point from Fukushima to Gunma," and "I did not bring my children along." This finding shows that today the Oze wetland management authorities must address concerns stemming from events outside the area as well.

15.4 Conclusion

In this brief chapter, I presented a historical overview of the Oze wetland and how it was recognized as a symbol of nature conservation in modern Japan, and how the value of the area was raised in the process. The Oze wetland has a distinct natural environment as it is a rare high-altitude wetland in Japan. Besides, the wetland is one of few remaining wetlands outside Hokkaido and houses many native species. However, the area is also easily accessible from Tokyo, and this has led to visitor pressure on the wetland. The chapter has shown how the calls for protection of the Oze area shaped the modern nature conservation movement of Japan through opposition to dam and roadways construction projects, and how the natural environment of the region became a catalyst for drawing together enlightened bureaucrats, alpinists, and conservation activists. The Oze model of national park management is currently followed in many other areas of Japan. However, the wetland itself remains afflicted by problems such as trampling of the vegetation, sewage treatment, invasive species, and overuse of resources. The natural environment and the human impact on Oze continue to attract public attention, due to the fact that the Oze wetland is widely recognized as a symbol of nature conservation in Japan.

References

Bansho K (2013) Change of the conservation consciousness and the conservation measures on Senjogahara Moor of Nikko National Park. Bull Univ Tokyo For 128:30–31 (In Japanese with English abstract)

Fukada K (1982) Nihon Hyakumeizan (The one hundred mountains of Japan). The Asahi Shimbun Company, Tokyo

Furuya K, Yui M, Akasaka M, Tada M, Ohata, T. (2001) Various problems of use under traffic control at Oze area in Nikko National Park. Tech Bull Fac Hortic Chiba Univ 55:21–41 (In Japanese with English abstract)

Geospatial Information Authority of Japan (2015) Research results of the area change of marshes and wetlands in Japan (Nihon Zenkoku no Shicchi Menseki Henka no Chosa Kekka). http://www.gsi.go.jp/kankyochiri/shicchimenseki2.html. Accessed 15 Apr 2017

Hatakeyama T (2001) Shizenhogohou Kougi (Lecture on the nature conservation laws). Hokkaido University Press, Sapporo (In Japanese)

Hirano C (1972) Oze ni Shisu (I shall die in Oze). Shinchosha, Tokyo (In Japanese)

Ishikawa T (2001) Nihon no Shizenhogo (Nature conservation of Japan). Heibonsha, Tokyo

Kikuchi K, Sudou S (1991) Towa no Oze (The eternal Oze) Jomo Shinbun, Maebashi, pp 150–152 (In Japanese)

MoE (Ministry of the Environment) (2016) Riyoushasuu Chousa (Survey of the visitor numbers). http://www.env.go.jp/park/oze/data/. Accessed 7 Apr 2017 (In Japanese)

Murakushi N (2005) Kokuritsukouen Seiritsushi no Kenkyu (Studies of the establishment of National Parks, Hosei). University Press, Tokyo

Nakajima K (1998) Oze Mondai to Kokuritsukoen Kanri (Issues of Oze and management of National Parks). Kokuritsu Koen (National Parks) 562:8–11 (In Japanese)

Natural Parks Foundation (2009) Nihon no Kokuritsukouen (National Parks of Japan). Natural Parks Foundation, Tokyo

Numata M (1994) Shizenhogo toiu Shisou (The concept of nature conservation). Iwanami Shoten, Tokyo

Oze Presevation Foundation (2002) Oze Shizen Kansatsu Gaido (Guidebook for nature observation). Yama-kei Publishers, Tokyo

Tamura T (1954) Shizenhogo Undou no Tenkai (The evolution of nature conservation movements in Japan). Kokuritsu Koen (National Parks) 61:2–6 (In Japanese)

The Asahi Shimbun (1957) Fumiarasareru Tennenkinenbutsu (Trampled natural monument). The Morning News Paper on June 9 (In Japanese)

The Asahi Shimbun (1965) Oze ni Hogoku (A project to establish a new protected area in Oze). The Morning News Paper on August 5 (In Japanese)

The Asahi Shimbun (1966) Meibutsu no Mokudo wo Fukusenka (Iconic Wooden walkways become multi-lane). The Evening News Paper on June 23 (In Japanese)

The Asahi Shimbun (1983) Oze no Kiseki (A miracle of Oze). The Evening News Paper on August 20 (In Japanese)

The Nature Conservation Society of Japan (NACS-J) (1985) Shizenhogo no Ayumi: Oze kara Tenjinzaki made (The history of nature conservation: from Oze to Tenjinzaki). The Nature Conservation Society of Japan, Tokyo (In Japanese)

The Nature Conservation Society of Japan (NACS-J) (2002) Shizenhogo NGO Hanseiki no Ayumi (The history of fifty years as nature conservation NGO). Heibonsha, Tokyo (In Japanese)

Yamamoto K, Kitabatake K (2012) Characteristics of the use and suggestions for the growth of use of visitor Centers in Oze National Park. Pap Environ Inf Sci 28:325–330 (In Japanese with English abstract)

16 Mount Fuji: The Volcano, the Heritage, and the Mountain

Abhik Chakraborty and Thomas E. Jones

Abstract

This chapter describes Mount Fuji, which at 3776 m asl forms the highest point in Japan, as a composite heritage. Fuji was registered in the UNESCO World Heritage Sites list in 2013 as a Cultural Property, but the mountain and its surrounding landscapes also represent a significant natural heritage. The steep, conically shaped mountain we see today almost wholly comprises volcanic products from the newest phase of Fuji volcanism, and a volcanic complex of three older volcanoes is buried under the ejecta from this new phase. Past eruptions of the mountain posed significant dangers for the densely populated Kanto plain; different landforms and ecosystems were formed out of those events that further underscore the mountain's value as a composite natural heritage. While Fuji is an exceptionally vigorous stratovolcano, many landforms in this area are formed by active denudation forces. Apart from the mountain, the surrounding areas offer a rich diversity of landforms such as artesian spring rivers, lava tunnels, and lakes. Tourism at Mount Fuji has changed significantly from its pilgrimage roots, and this transformation has put considerable pressure on the landscape. The natural heritage of this area urgently needs holistic planning and conservation measures to control the footprint of tourism and urban development.

Keywords

Mount Fuji • Stratovolcano • Composite heritage • Natural and cultural landscapes • Tourism

16.1 Introduction

Mount Fuji (Fig. 16.1), with its symmetrical conical shape, is an icon in the world of volcanoes. Rising 3776 m asl, the mountain forms the highest point of the Japanese Islands, and its height and shape make it a national symbol, a favorite topic for artists and writers, and an object of worship for the faithful (JNTO n.d.). The association of the mountain with national identity as well as local beliefs makes it an esthetic-religious symbol as well as a source of cultural sustenance, for which two reasons it was recognized as a UNESCO World Heritage (Cultural Property) (UNESCO n.d.). Located about 100 km west-southwest of Tokyo in the border between Shizuoka and Yamanashi Prefectures, Fuji is an active stratovolcano, which means it poses a significant hazard risk for the Tokyo conurbation (Yamamoto and Nakada 2015). Although Mount Fuji is described as a "solitary" volcano (UNESCO n.d.), the smooth conical shape of the mountain actually hides three older volcanoes; the Fuji we see today is in reality the result of accumulation of

A. Chakraborty (✉)
Center for Tourism Research (CTR), Wakayama University, Wakayama, Japan
e-mail: abhik78@hotmail.co.jp

T.E. Jones
Ritsumeikan Asia Pacific University, Beppu, Japan
e-mail: 110054tj@apu.ac.jp

A. Chakraborty et al. (eds.), *Natural Heritage of Japan*, Geoheritage, Geoparks and Geotourism,
DOI 10.1007/978-3-319-61896-8_16

Fig. 16.1 Mt. Fuji at evening light from Saiko Lake. Photograph courtesy Shamik Chakraborty

volcanic products in the latest phase of volcanism that can be related to the Fuji system. Thus, the mountain is not only a sacred object; it also possesses a special value for volcanologists and therefore is a significant part of the natural heritage of the Japanese Islands.

Due to its striking shape, height, and proximity to major urban centers, Mount Fuji is a popular tourist destination. Although the mountain was revered as a sacred space—the abode of the mountain goddess Konohana-Sakuyahime of the Fujisan Sengen faith (Hirano 1987)—today it is pressurized by urban and tourist infrastructure development, and by the sheer number of people who flock to the mountain every year. An estimated 300,000 people hike up the mountain every year (mainly during the summer climbing season that lasts from July to early September) (National Geographic Society n.d.), with 248,461 as the official number of hikers passing the 8th station in the 2016 season (MOE 2016). The UNESCO World Heritage Center document specifically mentions that the pressure brought by visitors and alteration of the landscape to cater for the demands of tourism affect both the integrity and the authenticity of the property (UNESCO n.d.). The case of Mount Fuji presents a significant challenge for conserving its natural and traditional cultural environments that together make it an object of outstanding universal value and an integral part of the natural heritage of the Japanese archipelago.

16.2 Fuji: The Composite Volcano and its Natural Heritage

16.2.1 A Brief Outline of Mount Fuji and its Predecessor Volcanoes

Mount Fuji began its activity about 100,000 years ago. Unlike most large stratovolcanoes of continental Japanese Islands, the main ejecta of Mount Fuji is basaltic in nature (Tsuchi 2007): at about 500 km^3, the deposits form the largest volcanic edifice in Japan (Miyaji et al. 2011). Takahashi (2000) has pointed out the positioning of Mount Fuji over a triple plate junction as the main reason of the upwelling of a large amount of tholeiitic basalt (see Fig. 16.2 for details of the location of Fuji). The term "Fuji volcanism" is usually associated with the volcanic products of Mount Ko-Fuji, which began its activity about 100,000 years ago, and Mount Shin-Fuji, whose main active phase started

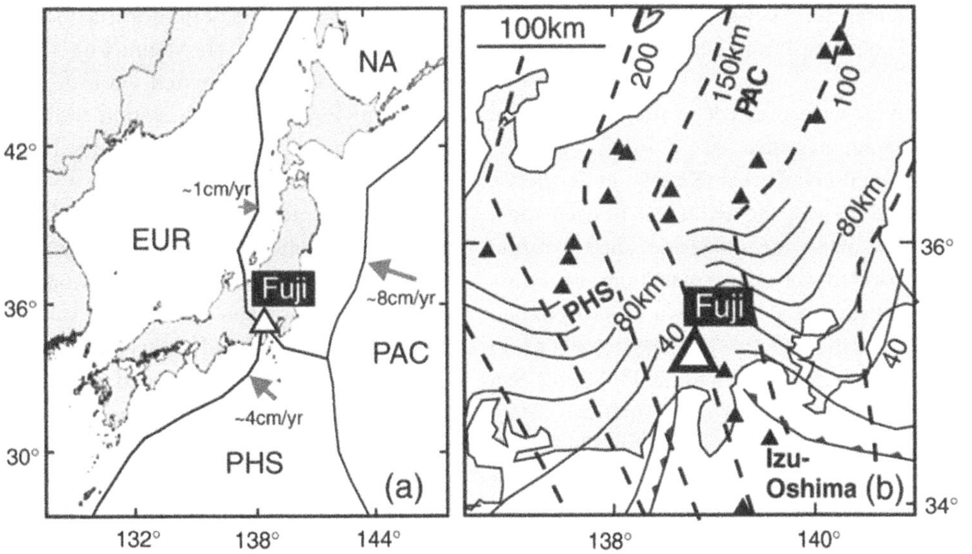

Fig. 16.2 Location of Mt. Fuji at subduction margin in the Japanese Islands. Interacting plates are EUR (European Plate, continental), NA (North American Plate, continental), PHS (Philippine Sea Plate, oceanic), and PAC (Pacific Plate, oceanic). Figure courtesy Kaneko et al. (2010). Reprinted from Journal of Volcanology and Geothermal Research, 193 (3–4) kaneko T, Yasuda A, Fujii T, Yoshomoto M., Cryptomagma chambers beneath Mt. Fuji, 161–170, Copyright (2010), with permission from Elsevier

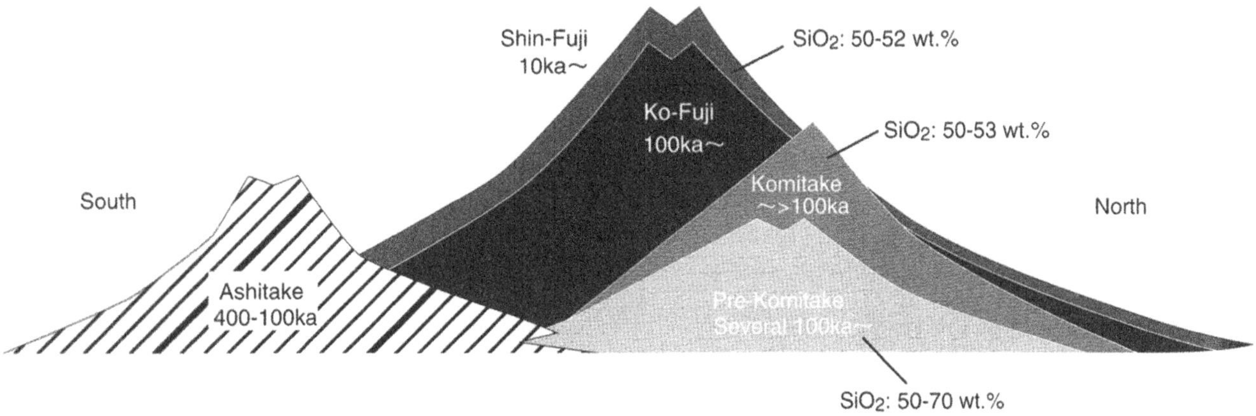

Fig. 16.3 Schematic diagram of the layered Fuji and Pre-Komitake volcanic systems (Courtesy Nakada et al. 2007, Reproduced with permission required for reproduction)

about 11,000 years ago (Ibid). According to Miyaji (2007), Shin-Fuji volcanism can itself be divided into five stages: continuous, very large-scale lava flows from the summit crater and flanks (17,000–8000 years ago); intermittent, small-scale explosive volcanism (8000–5600 years ago); intermittent, explosive volcanism with lava and pyroclastic flows from the summit crater and flanks (5600–3500 years ago); continuous but moderate explosive eruptions from the summit crater (3500–2200 years ago); and explosive volcanism with lava and pyroclastic flows from the flanks (2200 years ago-present).

However, while Shin-Fuji and Ko-Fuji are together referred to as "Fuji" volcanoes, an older volcano that ejected basaltic andesite was known to exist from an exposed section found on the northern slope of Mount Fuji (Shin-Fuji). This volcano is named Komitake and was thought to be the predecessor of Fuji volcanism (Machida 2007). However, more recent research has revealed that an even older volcano, Sen-Komitake (literally "Pre-Komitake") exists below the Komitake layer; it is now thought that this phase is the oldest phase of volcanism in the area (Nakada et al. 2007, see Fig. 16.3 above for a schematic diagram of the Fuji volcanic system). Nakada et al. (2007) provide a detailed explanation of the Pre-Fuji volcanism by focusing on the activities of Komitake and Sen-Komitake volcanoes which were active before 100,000 years ago and several hundred thousand years ago, respectively. While the discovery of an older volcano (Sen-Komitake) opened a new chapter in the history of volcanism in the Mount Fuji complex; many aspects of this exceptionally rapid and dynamic volcanic formation remain unclear; such as why the main craters have continuously shifted southward from Sen-Komitake to Shin-Fuji phases (Ibid).

In addition, due to the widespread deposition of Shin-Fuji products the volcanic systems of Ko-Fuji and its predecessors are not well understood (Kaneko et al. 2010). The basaltic magma of the volcano and traces of occasional explosive eruptions involving andesitic and dacitic components have inspired various explanations of Fuji's plumbing system. Nakajima and Hasegawa (2006) pointed out that the subducting slab of the Pacific Plate lies at a mean depth of 170 km below the mountain and the Philippine Sea Plate is under-thrust under the North American Plate (Kaneko et al. 2010). This geological setting led Fujii (2007) to postulate that the magma chamber of Fuji lies deeper than other volcanoes of the Izu-Bonin Arc, and the differences in the magma originates from high-pressure differentiation in this deep chamber. Kaneko et al. (2010) postulate that the volcano's plumbing system actually comprises two magma chambers—a "shallow" one (8–9 km below the surface) and a "deep" one (around 20 km below the surface). After each eruption, residual magma is left behind in the shallow chamber while basaltic magma rises from the deep chamber to mix with this residue before the following eruption. They also point out that sporadic explosive volcanism belongs almost exclusively to the newer phases of the Shin-Fuji volcano, which could suggest that although the eruption frequency of the volcano has decreased the possibility of explosive eruptions containing andesitic/dacitic magma has increased over the years.

16.2.2 Eruptions that Created Dynamic Heritage

The Shin-Fuji phase is mainly characterized by Strombolian eruptions; however, several past eruptions of Mount Fuji were known to be more destructive and far-reaching. While the summit crater has remained inactive since its last eruption about 2000 years ago, and subsequent eruptions have issued from its flanks (Miyaji 1988), the two most recent major eruptions (the Jogan Eruption of 864–66 and the Hoei Eruption of 1707–08) were particularly destructive (Miyaji 2007). The Hoei Eruption took place between 16 December 1707 and 1 January 1708 and is known for extensive ashfall up to Edo (Tokyo) (Miyaji et al. 2011). A detailed compilation from historical sources on the Hoei Eruption has been done by Koyama (2009). The Hoei event itself is a composite one—it has been subdivided into four phases—and the eruptions took place from three vents straddling 2100 m to 3000 m on the southeastern flank (Miyaji et al. 2011). Today, the remnant of the eruption—the Hoei craters and an elevated section on the southeast flank of Fuji formed by magma intrusion below the surface (called Hoei-zan or Mount Hoei)—form a tourist attraction along the Fujinomiya Trail on Mount Fuji's Western flank. The Hoei craters can be accessed via a detour 30 min into the hike from Fujinomiya 5th station trailhead; the largest of these craters has a diameter of nearly 1 km.

The Jogan eruption of 864–866 was comparatively less explosive but produced a similar volume of magma (Miyaji 2007). The eruption began in mid-June 864 AD, and the vent was located at a relatively low elevation on the northwestern flank of the volcano (Shizuoka University Center for Integrated Research and Education of Natural Hazards n. d.). The lava flow accumulated near the foothills of the northwestern flank; today, a dense forest that grew out of natural vegetation succession offers a fascinating example of a geo-ecological landscape. This 3000 ha. forest, known as the Aokigahara-Jukai, is so dense that travelers can lose their sense of direction once they venture deep into a landscape formed by large rocks and trees spreading their roots over them to form impassable barriers in places (Shizuoka University Center for Integrated Research and Education of Natural Hazards n.d.). The forests are an important habitat for wild birds as well as native vegetation. Visitors can also enjoy impressive lava tunnel formations at Narusawa Ice Cave and Fugaku Wind Cave. The Narusawa Ice Cave—a lava tunnel that formed when hot viscous lava flowed underneath a solidified upper crust and created a hollow tube-like formation—was designated as a natural monument in 1929. It offers breathtaking scenes such as ice columns, marks of the lava flow and lava tree molds[1] (Wind Cave and Ice Cave n.d.). The Ice Cave was used for preserving silkworm cocoons in the premodern era. The Fugaku Wind Cave is a short distance away from the Ice Cave and offers a similar spectacle, with impressive lava shelves and ropy pahoehoe lava formations (Ibid).

An even older eruption of the mountain of nearly 10,000 years ago left a geological heritage that also forms a lifeline for local societies: the Mishima Lava flow that issued from this large eruption was highly porous, and as it overlaid the older nonporous lava layer, it formed a gigantic pipeline for artesian water flow. The Kakitagawa Spring River is the most famous example of this "springwater." At 1.1 km in length, it is the shortest Class 1 River[2] in Japan, but still ranks among the largest springwater systems in the country in terms of the volume of springwater. The "river" is actually a combination of countless fissures that release artesian water flowing from the porous Mishima Lava aquifer. Apart from being an important habitat for aquatic vegetation

[1] Remnants of giant trees that were buried in the lava flow and were hollowed out.

[2] Japanese rivers are managed under two categories, Class 1 and Class 2. Class 1 rivers are typically larger and are considered more important for their environmental and water resource (utility) values.

and several fish and bird species, the Kakitagawa is the symbol of a citizens' movement against the depletion of groundwater through industrial overuse and construction. Established in 1988 by a group of local residents, the Kakitagawa Midori-no-Trust was one of the first national trusts in Japan (Kakitagawa Midori-no Trust n.d.).

Heritage in a Nutshell: Mt. Fuji

Location: Shizuoka and Yamanashi Prefectures, Central Honshu.

Access: Around 100 km from Tokyo, accessible either by Bullet Train to Shin-Fuji station (and local bus service thereafter to 5th station trailhead) or by direct Highway Bus from Tokyo.

Type of Heritage: Volcanic, composite. Peak Elevation 3776 m asl.

What to see: Mt. Fuji itself (including the summit crater and Hoei craters in the flank, volcanic ejecta), five lakes that are known for their scenic beauty and as picturesque settings to photograph the iconic volcano, waterfalls, spectacular lava cave and lava tube formations, the nearby Kakitagawa Spring River, and Shinto shrines belonging to the Fujisan faith.

Climbing Mt. Fuji: Nontechnical moderate hike from the 5th station trailheads (Hiking season is generally July to early September though exact dates differ slightly according to trails). There are four main trails: Yoshida, Fujinomiya, Subashiri, and Gotemba. Each trail offers a different view of the mountain and its surrounding landscape. Round-trip day-hike from the 5th station trailheads is possible, but staying overnight in one of the several mountain huts is also popular as it offers a chance to see the impressive sunrise from atop the mountain.

Other notes: Advisable to carry light hiking/trekking equipment (especially hiking shoes and rainwear), if climbing the mountain. If staying in mountain huts, pre-booking is advised. A 1000 JPY donation is requested of each climber at the trailhead to fund a variety of environmental conservation initiatives.

World Heritage Visitor Center: "Fujisan World Heritage Center": Located in Yamanashi Prefecture, Kawaguchi-ko Town. Open all year around, opening hours vary slightly according to seasons.

Useful websites:

http://www.fujisan-whc.jp/en/access/index.html

http://www.fujisan-climb.jp/en/

16.2.3 Other Notable Attractions in the Area

The summit crater of Mount Fuji is the main attraction for climbers and pilgrims. In premodern times, pilgrims used to take a round-trip of the summit crater, offering prayers to resident deities. The crater is 700 m in diameter and 200 m deep; a crater lake is thought to have existed about 1000 years ago (Yomiuri Shimbun and Koyama 2003). The bright red ejecta strewn all over the area is a scoria layer produced by the final eruptive phase of the summit crater 2200 years ago (Ibid).

Apart from the craters that impart the mountain's violent history, erosional landforms also offer important insights. For example, the Osawa Collapse is a deeply incised couloir that forms a gigantic gash 2100 m in length and 500 m wide at its widest point, with a depth of about 150 m (Yomiuri Shimbun and Koyama 2003). The Osawa Collapse offers a great cut-away view of lava layers below the surface, and it is a spectacular formation that speaks of the erosional forces that bring down mountains. Another erosion-generated feature is a "canyon" formed by torrents of snowmelt in warmer seasons that have cut through the soft pumice and scoria sections near the 5th station of the Subashiri trailhead (Ibid). Due to the sharp precipices and exposed layers of volcanic ejecta on both sides, the valley is popularly referred to as Mount Fuji's "grand canyon."

The five lakes around Mount Fuji form the biggest tourist attraction after the mountain itself. Located at the northern side of the mountain, these lakes are: Yamanakako (surface area of 6.57 km^2), Kawaguchiko (5.7 km^2), Saiko (2.1 km^2), Shojiko (0.5 km^2), and Motosuko (4.7 km^2) (Yamanashi Prefecture website n.d.). Known for their picturesque settings, the lakes are popular destinations for photographers. These water bodies were formed when lava flows from Mount Fuji blocked water outlets in valleys, and meltwater from the mountain, artesian flow, and rainwater accumulated over the years.

The Shiraito-no-Taki (literally, "White Thread Waterfall") is the largest waterfall near Mount Fuji; it has a vertical drop of 20 m and a width of 150 m (Fujinomiya City Website n.d.) and is a popular tourist attraction. The waterfall is located at the intersection of two lava layers, an older non-porous layer laid down by Ko-Fuji volcanism and a newer highly porous layer overlaid by Shin-Fuji volcanism—and the rich water supply is from snowmelt and groundwater discharge from the mountain (Yomiuri Shimbun and Koyama 2003).

16.3 Mount Fuji as a Tourist Landscape

Mount Fuji is the centerpiece of the Fuji-Hakone-Izu National Park, among the busiest in the world with over 110 million estimated annual visits per annum (MOE 2017). The majority of visits are clustered in the peak summer season that runs from July to mid-September, and attracts large number of climbers as well as holidaymakers.

Fuji's 5th station on the North side in Yamanashi attracts an estimated 3 million annual visitors, of whom some 10% attempt to reach the summit.

The summit of this stratovolcano has been revered as sacred since ancient times, and in the fourteenth century, Shugendo practitioners established a trail leading pilgrims to the summit (Polidor 2007). Thereafter, Fuji's peak became the goal for guided groups of pilgrim climbers, known as Fuji *kō*, who formed confraternities across Japan,[3] pooling contributions from its members to send representatives on a summer pilgrimage (Bernstein 2008). The climbing style differed significantly from current practice; selected climbers had ample opportunities to acclimatize during their pilgrimage, which could last a week or more. After the journey to Fuji, they stayed overnight at the foot before beginning an ascent from the base punctuated by numerous breaks en route to pray and perform ablutions.

To cater for the pilgrims, all the Fuji trailhead towns contained clusters of oshi inns[4] which formed networks of religious functionaries attached to shrines and temples around the foot of the volcano. As well as disseminating the Fuji-ko style of worship, they also provided "homes-stay" accommodation for the pilgrims. With the help of subordinate sendatsu,[5] the oshi organized logistics such as porters and provided spiritual services including blessings, rites, and ablutions. Pilgrims paid a package fee which included the devotional diet, purification and banners, as well as interpretation of local customs, thus ensuring a cost recovery mechanism for local communities. Their role as cultural mediators, combined with the substantial input for regional economies, has drawn comment on the parallels with current ecotourism practice (Ito 2009). Premodern pilgrim numbers are contested although they were estimated at over ten thousand in 1877 (Murakushi 2006).

During the post-war period of extended economic growth in Japan, climber and visitor numbers began to accelerate. Rising living standards brought both increased leisure time and improved access infrastructure, including the construction of many toll roads within national parks (Oyadomari 1989). At Fuji, the Subaru line toll road in Yamanashi was completed in time for the Tokyo Olympics in 1964. This 30 km paved road transports cars, buses, and bikes up to the Yoshida 5th station,[6] located at an altitude of 2300 m above sea level.

As well as the virtual disappearance of the religious dimension, another notable change in the climbing style has been the search for speed. The dramatic reduction in round-trip climbing time from the 5th station trailhead radically altered climbing behavior by transforming Fuji's summit into an overnight trek. Today, most climbers reach the summit via high-speed round-trips from Tokyo lasting less than 48 h, including a median 16–19 h spent on the climb itself (Jones and Yamamoto 2016). The ease of access for climbers from urban hubs has also facilitated a rise in footfall. There were around 100,000 summer climbers on the Yoshida trail in 1981.[7] The figure doubled to over 200,000 in 2008,[8] and since then the four trails have maintained a combined total[9] of approximately 300,000 climbers per season.

Each of the four main trails to the summit of Mt. Fuji has a 5th station trailhead that ranges in altitude from Gotemba (approx. 1400 m asl) to Fujinomiya (2400 m asl). Of the four, the Yoshida route on Fuji's north face of the mountain has the second highest elevation (2300 m asl), and its proximity to the Kanto plain ensures that it accounts for the majority of climbers: 60% of the total number of climbers on the mountain used the Yoshida trail in 2015 (MOE 2016). It was also reported to have the largest number of international climbers; a small but growing subsegment of the climber market (Jones et al. 2013).

Aside from the increasing volume of climbers, Fuji thus stands out among Japan's mountain destinations for having comparatively large numbers of young, inexperienced, first-time climbers—all factors with potential ramifications for risk management (Jones and Yamamoto 2016). Moreover, although the seasonal influx of 300,000 climbers in two months results in inputs to the regional economy; it also places considerable strain on the environment, particularly at management flashpoints such as toilets and trails. The provisioning of visitor facilities to lighten the load and mitigate impacts is thus of vital importance.

[3] By the early nineteenth century, there were 808 such groups in Tokyo alone (Bernstein 2008).

[4] Oshi were low-ranking Shinto priests. Their principal role was to disseminate the Fuji faith, but they also provided lodging for pilgrims in their homes which served as a base camp for climbing and to perform ablutions prior to the ascent (Kureha et al. 2015).

[5] Literally, "he who goes first," the Sendatsu led the mountain climbing and chanting.

[6] Each of the four main climbing routes comprises nine vertical "steps" divided up to Shugendo custom, with the tenth step the summit symbolizing heaven.

[7] The current manual counting system at the 6th station was introduced in 1981 along with a "Safety Centre" after 12 climbers were killed and 29 injured by a large-scale rock slide on 14 August 1980.

[8] The year in which Fuji was placed on the tentative list for inscription as UNESCO cultural heritage.

[9] As there is no system of climber registration, this data is estimated from visual monitoring by police (6th station) and infrared trail counters (8th station). It includes some double counts of staff, guides, repeat climbers, etc.

As pointed out above the quasi-religious pilgrim worship legacy, combined with an artistic heritage, formed the core components of a successful nomination application as a UNESCO cultural WHS in 2013. The recent increase in climber numbers and the WHS listing process prompted renewed debate on the issue of cost recovery mechanisms by management to mitigate such negative impacts. Although Japan's multipurpose nature parks have no entrance fee system, some visitor services come at a cost. For example, separate user fees are charged at the mountain huts, supplemented by the introduction in 1999 of a voluntary tipping system for visitors who use the toilets (Sayama and Nishida 2001). In lieu of the unfeasible compulsory entrance fees, donations collected from visitors could represent a new income source with which to provide services, and Shizuoka and Yamanashi prefectures co-initiated a new system in 2013 to raise funds for conservation. Although it began as a kind of "social experiment," the pilot scheme was introduced without much preamble, with few opportunities for a range of stakeholders to voice their opinions about the new system and the multiple issues associated with paying for outdoor recreation.

16.4 Issues for Heritage Conservation and Sustainable Area Management

Although Mount Fuji is registered in the World Heritage List as a cultural item, the area has significant natural heritage value as the preceding discussion shows. The single largest factor affecting the natural environment of Japan's most iconic mountain is the human footprint on the landscape in the form of tourist numbers and proliferation of infrastructure. The UNESCO website mentions that the integrity of the property remains a challenge in the face of urban development in the lower parts of the mountain, a large amount of structural intervention on the pilgrim route (alteration of trails by bulldozer roads to supply mountain huts, creating physical barriers to prevent movement of rocks, and the sheer number of travelers) interfere with the integrity of the mountain landscape (UNESCO 2016). In addition, UNESCO mentions the pressure of development in the lakeshore regions (especially Yamanaka-ko and Kawaguchi-ko), and the need for better integration of different natural and cultural components to project the value of this heritage is also mentioned (Ibid).

The authors have climbed Mount Fuji several times and visited the surrounding attractions in different seasons. We also witnessed tractors that hurtle down the trails and emit diesel fumes, as well as heavy machinery positioned at the summit. All can contribute significant negative impacts on the environment as well as the landscape. Mount Fuji today faces numerous tourism-related challenges; for example, the mountain huts are overcrowded during peak times in the climbing season, and thus the sense of solitude that is usually associated with pilgrimage cannot be felt easily. In addition, the surrounding areas are dotted with resorts and shops catering to the tourist market. In the case of the Kakitagawa Spring River, local conservation activists voiced their concern that only the 1.1 km long visible stretch of the river and its immediate surroundings are protected, while industrial water use and the felling of forests upstream have resulted in a significant reduction of the spring river's discharge over the years. Because the Kakitagawa is fed by an artesian mechanism that carries meltwater from the mountain nearly 30 km downstream, the lava layer along this whole 30 km stretch forms an "invisible river." Unfortunately, while the visible part of the stream—the 1.1 km stretch where it is above the surface—attracts attention, the river and its sustenance mechanism is not seen in its entirety. Such a lack of holistic vision also applies to different components of the mountain landscape. Machida (2007), for example, points out that a great amount of effort has been directed to "stop" the progression of couloir formation in the Osawa Collapse; but the geomorphological agents ensure that such structural intervention to alter the course of natural change has limited efficacy.

16.5 Conclusion

This chapter reviews Mount Fuji's role as "composite heritage," a term which refers not only to the combination of natural and cultural elements, but also underscores the fact that the geology of this area is highly dynamic and diverse. As described above, different phases of eruptions resulted in different landforms that in turn support different ecosystems. Ongoing denudation processes are a further component of land formation and landscaping; even as there is always a chance of an explosive eruption from the mountain. Although Mount Fuji has been extensively studied, scientists have so far only managed to categorically record the activity of the Shin-Fuji volcano, which is the newest phase of exceptionally vigorous volcanism spreading over several hundreds of thousands of years. The older phases of volcanism and their effect on landscapes and ecosystems remain largely obscure. Thus, Mount Fuji as a natural heritage requires further scientific research to understand land formation pathways, ecosystem formation, and progression of landscape features, and a holistic management scheme that can preserve the integrity of all of these components is required to protect this dynamic heritage.

Likewise, from a cultural heritage perspective, holistic management remains the goal. One encouraging example is the environmental conservation donation, a scheme jointly introduced by the two prefectures of Shizuoka and Yamanashi just after WHS inscription in 2013. Donations collected from visitors could generate new revenue streams with which to mitigate such negative impacts as trash and toilets. The new fund-raising scheme also answers calls from ICOMOS for improved stakeholder collaboration, bringing together the two adjacent prefectures Shizuoka and Yamanashi together with other actors from central and local governments, the private sector, and civil society. This cross-section of stakeholders must work together on the newly updated WHS strategy, a vital step toward a more holistic form of management.

Despite considerable changes in the climbing style and access arrangements, the pilgrim worship legacy remains one of the key storylines underpinning the successful 2013 listing as a World Heritage Site. More could be done to disseminate information about the pilgrims of yesteryear to the current crop of climbers, and marketing could be utilized to encourage climbers to make more sustainable choices in their itinerary. One promising example is a project piloted by Fuji Yoshida City to encourage climbers to ascend from the foot of the mountain, passing through the traditional Shinto shrine at the trailhead, and acclimatizing gradually to the altitude gain using the old path. By promoting this as the authentic climbing style, more climbers could be persuaded to climb from the bottom, thereby investing more time, effort, and money on their ascent.

References

Bernstein A (2008) Whose Fuji? religion, region, and state in the fight for a national symbol. Monumenta Nipponica 63(1):51–99

Fujii T (2007) Magmatology of Fuji Volcano. In: The Volcanological Society of Japan (ed) Fuji Volcano. Yamanashi Institute of Environmental Sciences, Fuji-Yoshida 233–244. (In Japanese)

Fujinomiya City (n.d.) Shiraito no Taki (Shiraito Waterfall). http://www.city.fujinomiya.shizuoka.jp/kankou/llti2b00000018ez.html. (In Japanese)

Hirano E (1987) Fujisan Sengen Shinko (Fujisan Sengen faith). Yuzankaku, Tokyo

Ito T (2009) Trail and climber management and its cost recovery from the climbers at Mt. Fuji in Edo Period. J Jpn For Soc 91(2):125–135

JNTO (n.d.) Mt Fuji Area, Shizuoka. https://www.jnto.go.jp/eng/regional/shizuoka/fujisan_shizuoka.html

Jones TE, Yamamoto K (2016) Segment-based monitoring of domestic and International climbers at Mount Fuji: targeted risk reduction strategies for existing and emerging visitor segments. J Outdoor Recreation Tour 13:10–17

Jones TE, Kato Y, Yamamoto K, Aramaki S (2013) The current state of international climbers at Mount Fuji: evidence from 2008-2012 monitoring surveys. J Fujiol 10(2):33–38

Kakitagawa Midori-no Trust (n.d.). http://www4.tokai.or.jp/kakita.rv-trust/. (In Japanese)

Kaneko T, Yasuda A, Fujii T, Yoshimoto M (2010) Crypto-magma chambers Beneath Mt. Fuji. J Volcanol Geothermal Res 193 (3–4):161–170

Koyama M (2009) Fujisan Funka to Hazado Mappu: Hoei Funka no 16 Nichikan (Eruption of MT Fuji and hazard map: a 16 day chronology of the Hoei eruption event). Kokon Shoin, Tokyo. (In Japanese)

Kureha M, Kikuchi T, Sano M, Yamamoto M (2015) Special issue: human geographical studies of Mount Fuji Region Focusing on Tourism. J Geogr (Chigaku Zasshi) 124(6):877–885

Machida H (2007) Development of Fuji Volcano: a review from quaternary tephrochronology. In: The Volcanological Society of Japan (Ed) Fuji Volcano. Yamanashi Institute of Environmental Sciences, Fuji-Yoshida, pp 29–44. (In Japanese)

Miyaji N (1988) History of Younger Fuji Volcano. J Geol Soc Jpn 94:433–452

Miyaji N (2007) Eruptive history eruption rate and scale of eruptions for the Fuji Volcano during the last 11000 years. In: The Volcanological Society of Japan (ed) Fuji Volcano. Yamanashi Institute of Environmental Sciences, Fuji-Yoshida, pp 79–95. (In Japanese)

Miyaji N, Kanno A, Kanamaru T, Mannnen K (2011) High-resolution reconstruction of the Hoei eruption (AD 1707) of Fuji Volcano, Japan. J Volcanol Geothermal Res 207(3–4):113–129

MOE (Ministry of the Environment) (2017) Shizen kouen no riyou shasuu (The number of annual visits to Japan's Nature Parks). http://www.env.go.jp/park/doc/data.html. Accessed 31 Jan 2017 (In Japanese)

MOE (Ministry of the Environment) (2016) Heisei 28 Nen Kaki no Fujisan Tozansha ni Tsuite (Data on Summer Hikers on Mt Fuji, 2016). http://kanto.env.go.jp/to_2016/28_3.html (In Japanese)

Murakushi N (2006) Kokuritsu kouen no seiritsu-shi (The history of the National Parks' establishment). Housei University, Tokyo, p 155

Nakada S, Yoshimoto M, Fujii T (2007) Pre-Fuji Volcanoes. In: The Volcanological Society of Japan (ed) Fuji Volcano. Yamanashi Institute of Environmental Sciences, Fuji-Yoshida, pp 69–77. (In Japanese)

Nakajima J, Hasegawa A (2006) Anomalous low-velocity zone and linera aignment of seismicity along it in the sub-ducted Pacific slab beneath Kanto Japan: reactivation of Sub-ducted Fracture Zone? Geophys Res Lett 33L:16309–16312

National Geographic Society (n.d.) Mount Fuji, Symbol of Japan. http://www.nationalgeographic.org/media/mount-fuji/

Oyadomari M (1989) The rise and fall of the nature conservation movement in Japan in relation to some cultural values. Environ Manage 13(1):23–33

Polidor A (2007) Sacred site reports. http://www.sacredland.org/mount-fuji. Accessed 11 July 2014

Sayama H, Nishida M (2001) Fujisan ni okeru sengo no bika seisou katsudou no hensen (Transition of beautification and clean-up activities at Mt. Fuji after World War II). J Jpn Inst Landsc Archit 64(5):485–488

Shimbun Y, Koyama M (2003) Katsu Kazan Fuji (Mt Fuji the active Volcamo). Chuo Koron Shinsha, Tokyo

Shizuoka University Center for Integrated Research and Education of Natural Hazards (n.d.). 3. Jogan Funka to Hoei Funka (3. Jogan and Hoei eruptions). http://sakuya.ed.shizuoka.ac.jp/sbosai/fuji/wakaru/003.html. (In Japanese)

Takahashi M (2000) Fuji Kazan to Maguma Kyokyu Sisutemu to Tekutonikusu ba: Mini Kakudai Kairyo Moderu (Magma supply system and tectonic system for Fuji Volcano: implications for mini-oceanic ridge expansion model). Gekkan Chikyuu (Chikyu Mon) 22:516–523. (In Japanese)

Tsuchi R (2007) Groundwater and springs of Fuji Volcano Japan. In: The Volcanological Society of Japan (ed) Fuji Volcano. Yamanashi Institute of Environmental Sciences, Fuji-Yoshida, pp 375–387. (In Japanese)

UNESCO (n.d.) Fujisan sacred place and source of artistic inspiration. http://whc.unesco.org/en/list/1418. Accessed 5 September 2016

Wind Cave & Ice Cave (n.d.) http://www.mtfuji-cave.com/en/

Yamamoto T, Nakada S (2015) Extreme volcanic risks 2: Mount Fuji. In: Papale P (ed) Volcanic hazards, risk and disasters. Elsevier, Amsterdam, pp 355–376

Yamanashi Prefecture (n.d.). Fujisan wo Torimaku Fuji Goko (The five lakes around Mt Fuji). https://www.pref.yamanashi.jp/japanese/tourism/mtfuji/lake.html. (In Japanese)

17 Synthesis: Toward Dynamic Conservation of Our Natural Heritage in the Anthropocene

Abhik Chakraborty

Abstract

This chapter provides a synthesis of the chapters of this volume. Based on the discussion provided in the foregoing chapters, it evokes the concept of "dynamic conservation" of natural heritage. The geological bearings of the landscape level and ecological aspects of the Japanese Islands' natural heritage are briefly revisited; and key discussion points of the case study chapters are summed up. It is shown how the chapters of this volume bring out the defining contours of the natural heritage of Japan, and how aspects of the dynamic earth revealed through these heritage landscapes could provide an important insight for heritage conservation based on the integral connections between abiotic and biotic processes that together make up natural heritage.

Keywords

Dynamic conservation • Anthropocene • Fragmentation • Earth processes • Geoecological turn

This book is not about providing a final word on natural heritage management; the chapters of this volume presented an outline of the varied and complex natural heritage of the Japanese Islands in order to stimulate our readers to think on how to evaluate and conserve them effectively. Accordingly our narrative focused on the "qualitative" aspects of such heritage, which have remained relatively underexplored in scientific studies of cause and effect relationships between phenomena. As discussed in Chap. 1 (Introduction), we live in an era of pervasive human influence in the geo-biosphere—the Anthropocene—that possibly implies that we are poised to undergo a radical departure in nature conservation in both theory and practice. The term "Natural Heritage" typically evokes perceptions of relatively unmodified places, where the integrity of geological, geomorphological, and ecological attributes is largely unaffected. But in today's world, reports are continuously emerging on fragmentation, failure to respond to aggressive degradation, and loss of heterogeneity in the natural heritage landscapes all over the world. Insufficient understanding of our natural heritage is possibly the most critical factor behind this state of affairs; and any nature conservation paradigm is examined for its capacity to address this incomplete understanding. What is paradoxical in this situation is that we already know a great deal about specific parts of our natural heritage, such as geology and ecology, and scientists such as Edward Wilson have repeatedly warned that the expansion of humanity's footprint has reached unprecedented and potentially calamitous level (Wilson 2013, 2016). Such warnings have come from the disciplines of economics, ecological and environmental studies as well; perhaps the most notable of which is the "Limits to Growth" argument (Meadows et al. 2004). Yet surprisingly little has been achieved in terms of conservation outcome on the ground, and there are fears that degradation of the natural environment, at least as far as biodiversity is concerned, is accelerating (Urban 2015). One reason behind this situation is the low level of integration of knowledge on

A. Chakraborty (✉)
Center for Tourism Research (CTR), Wakayama University, Wakayama, Japan
e-mail: abhik78@hotmail.co.jp

A. Chakraborty et al. (eds.), *Natural Heritage of Japan*, Geoheritage, Geoparks and Geotourism,
DOI 10.1007/978-3-319-61896-8_17

different aspects of the natural environment, the heritage of Planet Earth, and the lack of an interdisciplinary platform for active conservation. In this regard, combining the insights from geology, geomorphology, and ecology allows us to draw two important conclusions: (1) that "change" is a fundamental part of nature and that conservation science should benefit from explicitly referring to the earth processes that underpin all biodiversity and species habitats and (2) that conservation targets are not limited to individual species, landforms, or landscapes, but must incorporate the integrity of biophysical mechanisms acting over time. This is dynamic conservation; and it is needed for protecting our natural heritage in an age of accelerating anthropogenic change. What dynamic conservation implies is that there is no simplistic solution, nor are there simple targets for this type of conservation framework, and that it is based on the appreciation of the fundamental uncertainty and complexity of the natural systems (Stirling 2010). The Introduction chapter provided an outline of these issues.

Thereafter, we profiled the natural heritage of Japan in terms of its geological, geomorphological, and ecological characteristics and attempted to show the links between those attributes. In Chap. 2, it was discussed how the geological history of the Japanese Islands created a uniquely dynamic environment shaped by the interplay of multiple tectonic plates in an active margin. This geological history began nearly 750 Ma with the breakup of the supercontinent Rodinia and proceeded to unfold through tectonic events such as the formation of accretionary complexes and arc-arc collision due to a long history of subduction in the area. Such processes engendered an archipelago that is dominated by active volcanoes, a facet that was explored in detail in Chap. 3. In that chapter, it was also explained that although volcanoes are commonly seen as examples of nature's destructive force, those same forces produce significant ongoing landform and landscape diversity in the archipelago; and that the Japanese Islands are currently undergoing a quiescent phase in terms of intensity of volcanism.

Natural heritage case studies followed this thematic discussion. Natural heritage was grouped into three sections: UNESCO World Heritage, UNESCO Global Geoparks, and National Parks (selected sites). Of course, this division is for the ease of understanding; in reality, most WHS overlap with National Park territories and some geoparks also fall under National Park land.

In the UNESCO World Heritage (Natural) section, all four properties currently listed from Japan were covered. Each of those cases provides an instructive example of how geology, geomorphology, and the rich biotic diversity of those sites are firmly interrelated. Chapter 4 analyzed the Shiretoko Peninsula as a complex geoecological system that is shaped by oceanic and atmospheric circulation, faraway continental drainage, and is sustained by a land that has formed through rapid uplift and volcanism. Although the site is mostly well protected, trends such as the thinning of sea ice due to warmer winters and the fragmentation of riparian habitats because of damming and other types of artificial flow regulation pose continued questions for the integrity of the special ecosystems found here. It was argued that a reductive or species-specific conservation agenda is unlikely to work out, and the appraisal of Shiretoko as a complex, evolving system should be brought to the fore in planning and management schemes. The lesson is equally applicable to the next case study, that of the old-growth Siebold's Beech forest in the Shirakami Mountains (Chap. 5). The Shirakami area shows vigorous uplift and crustal deformation mechanisms; those mechanisms are fundamentally related to the rise and denudation of the mountains and formation of unique ecological niches. This chapter also touched upon the history of anthropogenic disturbance in the area—especially a planned forestry road in the 1980s that could have decimated this natural heritage—and argued for the vulnerability of this landscape, and for its intrinsic value. Chapter 6 presented the case of the Ogasawara Islands, which are known for their unique speciation history due to their remoteness from any continental landmass. Outlining the key ecological characteristics of the place, the chapter also outlined how invasive species and other types of anthropogenic pressure make these islands a vulnerable place. The final case study of this section (Chap. 7) analyzed the Yakushima Island, and especially the aspect of the relationship between the island's local community and the forest. The chapter also argued that while the World Heritage status results in increased recognition of the value of natural heritage, it could also paradoxically result in a simplification of value; thereby making certain aspects of the landscape obscure, while increased tourism could bring additional pressure on the heritage. As far as tourism is concerned, it can be concluded from all the four cases of this section that the sites represent a great opportunity for ecological education and ecotourism, but tourism in these areas must be carefully managed in order to protect sensitive ecosystems from further damage. While tourism can help generate valuable income for the local society, it typically results in intensification of resource use. For Natural World Heritage areas, it is certainly foreseeable that prudent planning, scientific analysis of tourism impact, and incorporation of local knowledge would make tourism a solution rather than a problem; but considering the vulnerability of certain aspects of the environment, restriction of access and creating no-visit zones are also important for elevating the value of natural heritage. This argument is bolstered by the fact that currently longitudinal data on changes at species, ecosystem, and landscape levels that can clarify causal relationships are scarce even in World

Heritage areas that are typically the best protected of all heritage landscapes.

In the second section, sites were selected from UNESCO Global Geoparks, which is a relatively recent heritage brand compared to the World Heritage. In Chap. 8, the history of how the geopark concept took root in Japan was briefly recounted, and how geoparks are evaluated for their earth heritage aspect was discussed. This chapter was followed by four case studies of geoparks: the San'in Kaigan, Muroto, Aso, and Unzen. While San'in Kaigan (Chap. 9) and Muroto (Chap. 10) feature tectonic themes of the back-arc spreading of the Sea of Japan (and the formation of the Japanese Islands as an archipelago away from the Asian mainland) and accretionary complex formation (due to ongoing plate subduction), respectively, Aso and Unzen are dominated by two of the best-known volcanoes of Japan. How geopark activities are complemented by local nature conservation was discussed in Chap. 9, while the Muroto case study showed how the geopark brand can raise local awareness for nature conservation. The Aso area (Chap. 11) features one of the largest grasslands of Japan formed in a caldera landscape created out of violent volcanic processes. It is an important biocultural landscape that represents many important Ecosystem Services (ES). The ES connection is a vital component for understanding pathways of change at the landscape level and forms a trusted toolkit for heritage conservation. The chapter also outlined the tight feedback loops between volcanism, grassland formation, and human use of landscapes of this area. Chapter 12 discussed the heritage of the Unzen volcano, especially in the light of the most recent eruption of Mt. Unzen. A valuable point for heritage management identified here is that for volcanic landscapes, recent activity typically "elevates" the heritage attraction to tourists; this lesson could be utilized for sustainable local development of these areas, which are periodically affected by economic loss due to volcanic eruptions.

Although geoparks are seen as a heritage brand that is useful for upholding the concept of the planet's abiotic diversity, as the analysis presented in Chap. 13 shows, geoparks currently suffer from a serious lack of conservation professionals, scientific conservation expertise, and inadequate monitoring of issues such as the fragmentation of geoheritage. Apparently, geoparks are focused more on the development agenda compared to the conservation agenda, and they lack a robust legal protection scheme or an explicit commitment to protect their signature sites. There is an urgent need for geoparks to provide tangible results of geoconservation. Perhaps increased interaction with international conservation initiatives and involvement of qualified professional could help to attain this goal. If geoparks can effectively address the issue of the ongoing fragmentation of the geo-biosphere and provide conservation results at the landscape level, they would certainly emerge as a key heritage label for protecting our natural heritage.

The third section featured case studies of Lake Akan National Park (Chap. 14), Oze National Park (Chap. 15), and Mt. Fuji (Chap. 16). These national parks possess important aspects of natural heritage: in the case of Akan, it is the presence of the marimo algae colony in a lake formed by a long sequence of caldera forming eruptions; in case of Oze, it is the highland marsh which incidentally became a major starting point of the nature conservation movement in Japan; and in case of Mt. Fuji, it is a complex volcanic history that has resulted in a landscape of great beauty and cultural significance, in addition to its significant role in ongoing land formation.

Together, these case study chapters bring out the defining contours of the natural heritage in Japan. While the Japanese Islands are relatively young in geological terms, for this very reason, they provide an excellent setting for observing the geological-geomorphological-ecological linkages, as much of the archipelago is still undergoing significant landformation events. Ongoing earth processes of tectonic movement, volcanism, erosion, and their feedback to the biological world endow this area with potential for grounding a "geoecological" turn in natural heritage conservation. It is quite evident from the discussion provided in foregoing chapters that conserving such multifaceted and diverse heritage requires the understanding of the planet's processes operating over vast scales of space and time. In this regard, the natural heritage of Japan is not only confined to its geographical boundaries, but is a heritage that provides a snapshot of the life history of our planet.

References

Meadows DH, Randers J, Meadows G (2004) Limits to growth: the 30 year update. Chelsea Green Publishing, Chelsea, VT

Stirling A (2010) Keep it complex. Nature 468:1029–1031

Urban MC (2015) Accelerating extinction risk from climate change. Science 348(6234):571–573

Wilson EO (2013) The social conquest of Earth. Liveright (WW Norton & Co.), New York

Wilson EO (2016) Half Earth: our planet's fight for life. Liveright (WW Norton & Co.), New York

Index

A. Chakraborty et al. (eds.), *Natural Heritage of Japan*, Geoheritage, Geoparks and Geotourism,
DOI 10.1007/978-3-319-61896-8